FOOD SCIENCE AND TECHNOLOGY

THE FEDERAL FOOD SAFETY WORKING GROUP

PROGRESS AND ASSESSMENTS

FOOD SCIENCE AND TECHNOLOGY

Additional books in this series can be found on Nova's website under the Series tab.

Additional e-books in this series can be found on Nova's website under the e-book tab.

FOOD SCIENCE AND TECHNOLOGY

THE FEDERAL FOOD SAFETY WORKING GROUP

PROGRESS AND ASSESSMENTS

RICHARD S. JAYNES
AND
SELENA DARLINGTON
EDITORS

New York

Library of Congress Cataloging-in-Publication Data

ISBN: 978-1-62417-059-1

Published by Nova Science Publishers, Inc. † New York

CONTENTS

PREFACE

American consumers deserve to trust the safety of the food they purchase for themselves and their families. While we enjoy one of the safest food supplies in the world, it requires constant monitoring. We need to be able to rapidly identify and address risks to our food supply as a result of new disease agents, new food technologies, changes in U.S. demographic and dietary patterns, and an abundance of food imports resulting from an increasingly globalized food supply. The federal government has taken important steps to protect an increasingly complex food supply. In March 2009, President Obama created the Food Safety Working Group (FSWG), a central coordinating mechanism for the federal government's food safety activities that is led by the Department of Health and Human Services and the USDA. This book provides an overview of the progress and assessments of the Federal Food Safety Working Group with a focus on results oriented goals and performance measures for food safety oversight.

Chapter 1 – The premise underlying the FSWG's creation is that coordination among federal agencies is essential to protect consumers in our highly diverse, global food system. Coordination and cooperation are necessary through every stage of the production and consumption process—"from farm to table." FSWG member departments and agencies thus share information and experience about all aspects of food safety. Such coordination strengthens the scientific and technical infrastructure to support a modern food safety system.

Collaboration with state, local and foreign governments as well as partnership with the private sector is likewise crucial.

The FSWG recognizes that through enhanced collaboration and selective, risk-based deployment of government efforts, the federal agencies can best utilize all available food safety resources to ensure the effectiveness and efficiency of the food safety system and improve food safety programs.

Chapter 2 – For more than a decade, GAO has reported on the fragmented nature of federal food safety oversight and how it results in inconsistent oversight, ineffective coordination, and inefficient use of resources. In 2007, GAO added this issue to its high-risk list. In March 2009, the President established the Food Safety Working Group (FSWG) to coordinate federal efforts and establish food safety goals to make food safer.

Section 21 of Public Law 111-139 mandated that GAO identify programs, agencies, offices, and initiatives with duplicative goals and activities. This review examines: (1) steps, if any, that the FSWG has taken to increase collaboration among federal food safety agencies, and (2) options we and others have identified to reduce fragmentation, overlap, and potential duplication in food safety oversight. GAO reviewed information about the FSWG and alternative organizational structures for food safety, and conducted interviews.

Chapter 3 - Numerous federal, state, and local agencies share responsibilities for regulating the safety of the U.S. food supply. Federal responsibility for food safety rests primarily with the Food and Drug Administration (FDA) and the U.S. Department of Agriculture (USDA). FDA, an agency of the Department of Health and Human Services, is responsible for ensuring the safety of all domestic and imported food products (except for most meats and poultry). FDA also has oversight of all seafood, fish, and shellfish products. USDA's Food Safety and Inspection Service (FSIS) regulates most meat and poultry and some egg products. State and local food safety authorities collaborate with federal agencies for inspection and other food safety functions, and they regulate retail food establishments. The combined efforts of the food industry and government regulatory agencies often are credited with making the U.S. food supply among the safest in the world. However, critics view this system as lacking the organization, regulatory tools, and resources to adequately combat foodborne illness—as evidenced by a series of widely publicized food safety problems, including concerns about adulterated food and food ingredient imports, and illnesses linked to various types of fresh produce, to peanut products, and to some meat and poultry products. Some critics also note that the organizational complexity of the U.S. food safety system as well as trends in U.S. food markets—for example, increasing imports as a share of U.S. food consumptions and increasing consumption of fresh, often unprocessed, foods—pose ongoing challenges to ensuring food safety. The 111th Congress passed comprehensive food safety legislation with the FDA Food Safety Modernization Act (FSMA, P.L. 111-353). FSMA is the largest expansion of FDA's food safety authorities since the 1930s. Although numerous agencies share responsibility for regulating food safety, FSMA focused on foods regulated by FDA and amended FDA's existing structure and authorities, and did not directly address meat and poultry products under USDA's jurisdiction. Beyond these changes, some in Congress continue to push for additional policy reforms to address other perceived concerns about the safety of the U.S. food supply.

In: The Federal Food Safety Working Group ISBN: 978-1-62417-059-1
Editors: R. Jaynes and S. Darlington

Chapter 1

THE FEDERAL FOOD SAFETY WORKING GROUP PROGRESS REPORT*

The Federal Food Safety Working Group

INTRODUCTION

American consumers deserve to trust the safety of the food they purchase for themselves and their families. While we enjoy one of the safest food supplies in the world, it requires constant monitoring. We need to be able to rapidly identify and address risks to our food supply as a result of new disease agents, new food technologies, changes in U.S. demographic and dietary patterns, and an abundance of food imports resulting from an increasingly globalized food supply. In recent years, consumers and industry alike have been impacted by illnesses associated with food products, such as ground beef, peppers, peanut butter, spinach, shell eggs, and cookie dough, among others. While regulatory and industry efforts have over time improved food safety considerably, the Centers for Disease Control and Prevention has recently estimated that 1 in 6 Americans suffers from foodborne illness annually, resulting in 128,000 hospitalizations and 3,000 deaths per year, most of which are preventable.

The public health consequences of foodborne illness remain significant. The impact reaches well beyond the number of cases. Foodborne illnesses result in billions of dollars in medical costs, as well as significant economic

* This is an edited, reformatted and augmented version of The Federal Food Safety Working Group Progress Report, dated December 2011.

losses to the food industry when illness outbreaks and contamination incidents undermine consumer confidence in affected commodities, require large recalls, and diminish demand. American consumers have high expectations for the safety of the food supply. While all risks will never be eliminated, the public rightfully expects that the government and the food industry do everything that can reasonably be done to prevent food safety problems.

The federal government has taken important steps to protect an increasingly complex food supply. In March 2009, President Obama created the Food Safety Working Group (FSWG), a central coordinating mechanism for the federal government's food safety activities that is led by the Department of Health and Human Services (HHS) and the U.S. Department of Agriculture (USDA). The FSWG's core mission is to strengthen federal efforts and develop short-term and long-term strategies to improve food safety. Partner agencies include HHS's Food and Drug Administration (FDA), USDA's Food Safety and Inspection Service (FSIS), and HHS's Centers for Disease Control and Prevention (CDC), as well as the Environmental Protection Agency (EPA), Department of Homeland Security, Department of Commerce, Department of State, and the Office of the United States Trade Representative. The White House Domestic Policy Council convenes the FSWG.

The premise underlying the FSWG's creation is that coordination among federal agencies is essential to protect consumers in our highly diverse, global food system. Coordination and cooperation are necessary through every stage of the production and consumption process—"from farm to table." FSWG member departments and agencies thus share information and experience about all aspects of food safety. Such coordination strengthens the scientific and technical infrastructure to support a modern food safety system.

Collaboration with state, local and foreign governments as well as partnership with the private sector is likewise crucial.

The FSWG recognizes that through enhanced collaboration and selective, risk-based deployment of government efforts, the federal agencies can best utilize all available food safety resources to ensure the effectiveness and efficiency of the food safety system and improve food safety programs.

By clarifying responsibilities and improving accountability, the FSWG has already strengthened the nation's food safety system. The FDA has created a new position, Deputy Commissioner for Foods, who is empowered to restructure and revitalize FDA's work developing a new food safety system.

The FSWG agencies have also taken substantial steps to modernize federal food safety. For instance, FSIS launched the Public Health Information

System (PHIS), a comprehensive data analytics system to better detect and respond to foodborne hazards. FDA has increased the number of domestic and foreign risk-based inspections it undertakes and is pursuing a comprehensive, farm-to-table strategy for preventing food safety problems.

CDC has begun development of next-generation laboratory methods and new epidemiological tools for investigating multi-state outbreaks in partnership with State, local, tribal, and territorial health departments.

The following pages detail some of the FSWG's other core efforts and accomplishments over the past two years to strengthen the food safety system through greater prevention, surveillance, and response.

In July 2009, the FSWG submitted a Report of Key Findings to the President. That report emphasized a three-dimensional approach to enhancing public health through greater food safety:

- **"Prevention"**— establish science-based best practices to reduce the risk of illness among entities that produce, process, and distribute food;
- **"Surveillance"**— the ongoing, systematic collection and analysis of containment, public health, and molecular data, throughout the farm-to-fork continuum, for use in preventing and controlling foodborne illnesses; and
- **"Response"**— rapidly detect and terminate foodborne illness outbreaks and contamination when and where they do occur.

Over the past two years, these aims have provided the organizing framework for the FSWG's efforts, and they will continue to do so going forward.

Two years after its creation, the time is right to assess the FSWG's efforts to date and to anticipate its agenda over the next two years. Accordingly, this Progress Report summarizes some of the FSWG's accomplishments to improve prevention, surveillance, and response. It also identifies some of the FSWG's priorities over the next year and beyond, including activities necessary to implement the FDA Food Safety Modernization Act (FSMA), which was signed into law on January 4, 2011.

The passage of FSMA is one of the most significant accomplishments in the history of food safety in the United States. It was passed on a bi-partisan basis with support from a broad coalition of industry, consumer and public health groups and is grounded in the same principles that the FSWG embraced in its 2009 report.

FSMA calls on FDA, working in partnership with other federal, state and local agencies and the food industry, to build a modern new system of food safety oversight that harnesses the best available practices to prevent food safety problems. FSMA also calls on CDC to strengthen public health surveillance and response networks in partnership with state and local health departments. Implementation of FSMA will be a major FSWG priority in the years ahead.

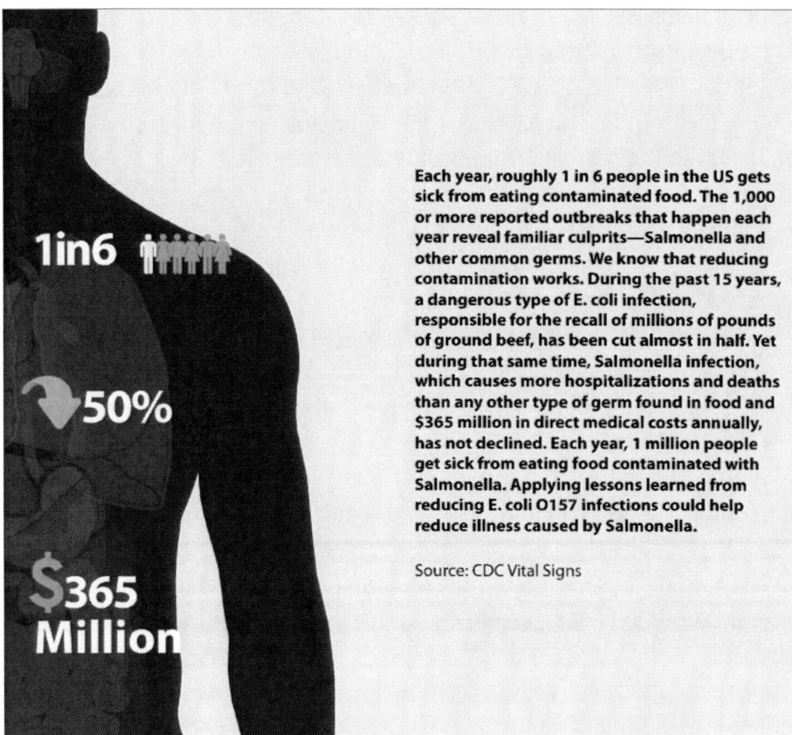

I. Food Safety Working Group Initiatives and Accomplishments

A. Prevention of Foodborne Illness

Prevention is the front line of food safety defense. The FSWG agencies have taken substantial steps to reduce the occurrence of foodborne illness by preventing the contamination and adulteration of foods. They have focused on *Salmonella, E. coli, Campylobacter,* and *Listeria* among other causes of foodborne illness.

1. Reduce Bacterial Pathogens in Foods

Salmonella and *Campylobacter* are the most frequently reported causes of foodborne illness, and both FDA and FSIS have focused on efforts to combat these two pathogens. *Salmonella* causes an estimated one million foodborne illnesses each year and, for more than a decade, eggs have been identified as a leading cause of *Salmonella* illnesses. Yet, despite support from consumer advocates and the egg industry, the federal government had not established basic rules on egg safety to prevent contamination. In July 2009, FDA issued an egg safety rule to control *Salmonella* Enteritidis contamination of shell eggs during production. This rule is expected to reduce the number of foodborne illnesses associated with consumption of raw or undercooked contaminated shell eggs by approximately 60 percent, an estimated 79,000 illnesses every year, and to generate savings of nearly $1 billion in healthcare costs and costs to the industry per year. The requirements of the rule went into effect for the largest egg producers in July 2010. To ensure compliance with the rule, FDA will, by the end of 2011, inspect 600 of the largest producers, who account for 80 percent of the U.S. shell egg supply. To ensure the safety of liquid eggs as well, USDA and FSIS have undertaken an "Egg Risk Assessment," and a baseline study on liquid egg products began in the fall of 2011.

In August 2010, FDA published a notice of availability of a Draft Guidance for Salmonella in Animal Feed. This guidance provides clarity to regulatory officials and animal feed industry on when the presence of *Salmonella* in animal feed causes the feed to be adulterated.

In addition, FSIS has taken further action to prevent illness by implementing stricter pathogen reduction performance standards for *Salmonella.* These standards aim to reduce the presence of *Salmonella* in young chicken (broiler) and turkey slaughter establishments.

FSIS will publish the names of failing establishments on its website to inform the public of risks to food safety. The agency has the goal of having 90 percent of poultry establishments meet or exceed the revised *Salmonella* standards.

FSIS also implemented its first-ever pathogen reduction performance standards for *Campylobacter* in poultry establishments. Revising current performance standards—and setting new ones—will continue the reduction of pathogens and result in safer products.

In addition, FSIS announced that it will expand the Salmonella Initiative Program (SIP) to help reduce *Salmonella* in raw meat and poultry products, a program that aims to reduce and eliminate pathogens before products reach consumers. This voluntary, incentive-based program will allow participating establishments to operate under certain regulatory waivers in order to try new procedures, equipment or processing techniques to better control *Salmonella.* Establishments under SIP will collect product samples, use these samples to test for common foodborne pathogens such as *Salmonella, Campylobacter* and *E. coli,* and share this internal food safety data with FSIS. This approach improves food safety by encouraging industry to test for and take action to reduce pathogens.

E. Coli O157:H7

The bacterial strain *E. coli* O157:H7 causes diarrhea, abdominal pain, and fever in approximately 63,000 Americans each year following consumption of contaminated food. In an estimated one in fifteen patients, a serious complication known as "Hemolytic-Uremic Syndrome" develops. Patients with this complication can suffer intense pain, anemia, kidney failure, and even death. In recent years, this bacterium has caused outbreaks associated with meat and spinach, among other foods. In March 2010, FSIS reissued a directive and began a new verification testing program for beef bench trim (a key cut of meat used in making ground beef) and for establishments that handle beef, to ensure they are taking action to reduce the presence of *E. coli* O157:H7. The directive includes instructions to both domestic and important inspection personnel for sampling raw beef products for FSIS verification testing for *E. coli* O157:H7.

In addition, it outlines actions that FSIS will take if samples of raw ground beef, raw ground beef components, or raw beef patty components test positive for *E. coli* O157:H7. Finally, it includes instructions for other verification activities concerning *E. coli* O157:H7. FSIS has also increased their sampling to find this pathogen, focusing largely on the components that go into making ground beef. Foodborne Diseases Active Surveillance Network's ("FoodNet") 2010 data indicates that the substantial decrease in incidence of reported *E. coli* O157:H7 infections met the national Healthy People objective for 2010, and a further decrease of 50% is called for to meet the 2020 national objective.

E. Coli Non-O157

FSIS announced that six additional serogroups of pathogenic *E. coli* will be declared adulterants in non-intact raw beef—raw ground beef, its components, and tenderized steaks. These products will be prohibited from sale to consumers, and FSIS is launching a testing program to detect these dangerous pathogens and prevent them from reaching consumers. CDC estimates that over 112,000 foodborne illnesses occur every year from non-O157 STEC and that the six new *E. coli* serogroups (026, 0103, 045, 0111, 0121 and O145) are responsible for the greatest number of non-O157 Shiga toxin-producing *E. coli* (STEC) illnesses, hospitalizations and deaths in the U.S. *E. coli* O157:H7 already is an adulterant in these products.

Listeria

The bacterium *Listeria monocytogenes* poses a risk to our most vulnerable populations—children, the elderly, and pregnant women— and exhibits a high mortality rate in those infected. A joint interagency risk assessment involving FSIS and CDC on *Listeria monocytogenes* contamination in retail facilities is underway through the application of a cross-contamination model. This risk assessment will inform decision-making about how to best address this pathogen in the retail setting, such as in deli counters at supermarkets. Moreover, there is significant scientific research sponsored by the food safety agencies that will help elucidate how to best prevent *Listeria* contamination at retail.

2. Improved Produce Safety

Salmonella and *E. coli* O157:H7 are also threats to the safety of fresh produce. According to FDA data, between 1996 and 2010, at least 131 foodborne illness outbreaks were associated with the consumption of fresh produce.

In July 2009, building on experience gained through collaborative efforts with the produce industry, the FDA issued commodity-specific draft guidance on agricultural practices that can reduce the risk of microbial contamination in the production and distribution of tomatoes, melons, and leafy greens — three commodities that pose higher risk of such contamination than do other commodities. That draft guidance was intended to assist firms by recommending practices and preventive measures to minimize the microbial food safety hazards of the products throughout the entire supply chain and thereby create a minimum standard for production across the country.

FDA also committed in 2009 to establishing mandatory standards for safe growing practices that would build on its commodity-specific draft guidance and take account of the full diversity of the produce sector in terms of risk, growing practices and scale of operation. FDA has held listening sessions in thirteen states across the country to gather input for a new, comprehensive proposed rule to reduce the risk of foodborne illnesses from the consumption of fresh produce through the implementation of preventive controls. Preventive controls for the produce industry will provide a consistent set of standards for the industry and reduce the risk of consuming contaminated produce.

The Produce Safety Alliance, a three-year public-private partnership funded by FDA and USDA and housed at Cornell University, was established to help produce growers and packers access food safety education materials. This is a key step in preparation for new science-based standards for produce safety that are required by the Food Safety Modernization Act. Key activities of the alliance include: (1) Developing a standardized, but multi-formatted and multi-lingual education program on Good Agricultural Practices and co-management, (2) Creating an information bank of up-to-date scientific and technical information related to on-farm and packinghouse produce safety, (3) Launching a website to make the alliance's work and information readily accessible, and (4) Establishing a network of educational collaborators.

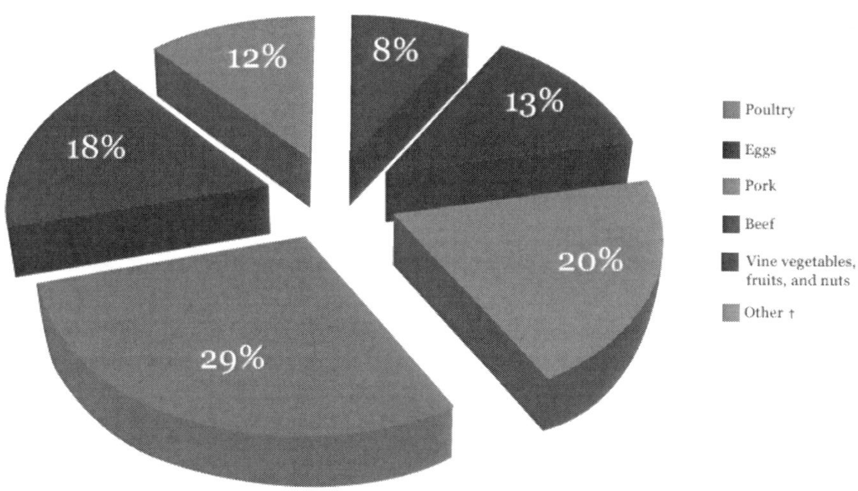

*These contaminated ingredients or single foods (belonging to one food category) were associated with 1/3 of the *Salmonella* outbreaks.

†Other includes: Sprouts, leafy greens, roots, fish, grains-beans, shellfish, oil-sugar, and dairy.

Source: CDC National Outbreak Reporting System, 2004–2008.

Foods associated with *Salmonella* outbreaks.*

3. Preventing Intentional Adulteration

FSIS, FDA, and DHS collaborated with the National Center for Food Protection and Defense (NCFPD) to study economically motivated adulteration (EMA) of food products. Several potential EMA indicators were identified based on the information obtained. Quantitative measures are being developed to help identify potential EMA incidents.

FDA has also been working on several other fronts to improve food defense. First, FDA is working with major trade partners to improve global capacity to prevent intentional contamination and protect imports. Second, FDA is creating tools including: (a) vulnerability assessment software — agriculture and manufacturing modules; (b) mitigation strategies database—searchable by process step; (c) exercise kit for use by State and local public health departments to prepare for foodborne outbreaks, both unintentional and intentional. Third, FDA is conducting vulnerability assessments, using industry and academic expertise to continue to build our understanding of food systems to inform future efforts.

4. Other Preventive Measures in Food Safety

FSIS has issued several instructions to its inspection program personnel to improve the safety of meat and poultry produced at regulated establishments.

Residues

Residual drugs, pesticides, and environmental contaminants such as metals that are sometimes found in meat and poultry can pose a public health risk when found at certain levels. The National Residue Program (NRP), administered by FSIS, FDA, and EPA, is in place to protect the public from these residues. A newly formed Senior Executive Council has been charged with establishing Standard Operating Procedures (SOP) to review and improve the protocols of the NRP's sampling program, but foremost, to determine appropriate regulatory actions for cancelled pesticides and environmental contaminant residues. FSIS has also strengthened its enforcement activities by instructing its inspection program personnel to increase chemical residue testing in certain slaughter establishments that do not have an effective residue control program in place.

Allergens

In the first half of 2011, an increased number of products were recalled because of undeclared allergens or other ingredients, a significant public health concern due to the rising number of Americans sensitive to ingredients, such as milk, eggs, peanuts, tree nuts, soy, wheat, seafood and shellfish, that elicit potentially life-threatening responses.

FSIS inspectors were instructed to make establishments aware of the importance and prevalence of undeclared allergens in meat and poultry products, and how to best ensure labels are accurate and current.

Test and Hold

FSIS proposed a new requirement for the meat and poultry industry that, once enacted, will reduce the amount of unsafe food that reaches store shelves. With the proposed requirement, FSIS would be able to hold products from commerce until FSIS test results for harmful substances—such as *E. coli* O157:H7 and drug residues—are received. Currently, when FSIS collects a sample for testing, the sampled products are requested but not required to be held until test results are known. FSIS believes that this requirement will substantially reduce recalls for meat and poultry products.

Infant Formula

FDA is finalizing its rule on infant formula, which will establish new or updated requirements including the following: (1) current good manufacturing practices, including audits, (2) growth monitoring studies to ensure that infant formulas support healthy growth, (3) microbial testing for pathogens of concern and (4) recordkeeping.

B. Enhanced Food Safety Surveillance and Compliance

In addition to supporting greater efforts to prevent foodborne illness in the first place, the FSWG has also taken important steps to strengthen the capacity of the food safety system to monitor the cause of illnesses, detect emergent outbreaks as early as possible, and ensure compliance with food safety standards, with a special focus on improving oversight of imports.

1. Disease Surveillance

CDC has been working with state and local partners to achieve a comprehensive approach for foodborne illness surveillance that detects outbreaks; estimates illnesses, hospitalizations, and deaths; determines foods and settings causing illness; and tracks trends to determine whether control measures are working. For example, CDC works with public health partners to enhance the public health surveillance network for outbreak detection. Specifically,

PulseNet, a national network of public health and food regulatory agency laboratories which is coordinated by the CDC, is now in all 50 states and 82 countries. PulseNet participants perform standardized molecular subtyping (or "fingerprinting") of foodborne disease-causing bacteria by pulsed-field gel electrophoresis (PFGE). PFGE can be used to distinguish strains of organisms, such as Escherichia coli O157:H7, Salmonella, Shigella, Listeria, or Campylobacter at the DNA level. PulseNet allows for rapid detection of foodborne disease case clusters and real-time communication among state, local health departments, and international partners, as well as facilitates early identification of common source outbreaks.

Additionally, the CaliciNet surveillance system for noroviruses is now operational in 25 states. In August 2010, CDC expanded FoodCORE (Foodborne Diseases Centers for Outbreak Repsonse Enhancement) following an initial pilot program, which was supported in part by FSIS. Seven sites are implementing work plans to build replicative models for enhanced outbreak response activities. Resources have already played a critical role in the rapid containment of several recent outbreaks, both single- and multi-state, including *E. coli* O157 and hazelnuts, and *Salmonella* in Lebanon bologna, chicken livers and queso fresco. FoodCORE resources facilitated rapid and comprehensive laboratory, epidemiologic, and environmental health response during these investigations.

CDC is meeting its obligations under the Food Safety and Modernization Act (FSMA) by establishing a Surveillance Working Group that consists of members of federal, state and local governments, academia, industry, and consumer groups. The Working Group met and offered guidance related to the designation of Food Safety Integrated Centers of Excellence, also mandated by FSMA. Future Working Group efforts will be focused on enhancements to foodborne diseases surveillance.

CDC launched a new web-based surveillance platform to enhance the speed and completeness of foodborne outbreak reports. CDC also developed an online database to make data more publicly accessible and better inform both detection and prevention efforts. In January 2011, CDC published revised and updated estimates of foodborne illnesses hospitalizations and deaths in the United States, which form the basis for future policies and research in food safety. In June, 2011, CDC published the FoodNet annual report card on food safety in the United States, showing that *E. coli* O157 infections have been reduced, but *Salmonella* infections have not declined in 15 years.

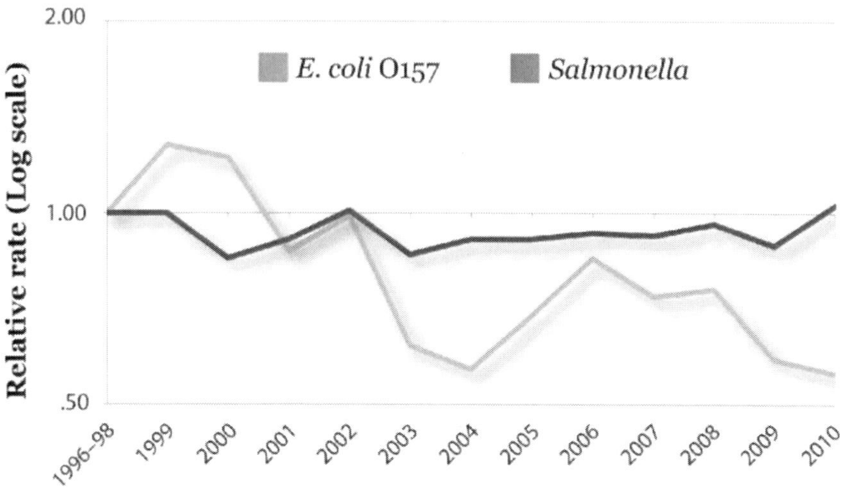

Source: Foodborne Diseases Active Surveillance Network, 2010.

Change in E.coli O157 and *Salmonella* infection, 1996-2010.

2. Reportable Food Registry

In September 2009, the FDA launched the congressionally-mandated electronic Reportable Food Registry (RFR) portal for industry and public health officials to report significant food safety incidents, such as contamination of food with pathogenic bacteria. Specifically, registered food facilities that manufacture, process, pack, or hold food for human or animal consumption in the United States under section 415(a) of the Food, Drug and Cosmetic Act are required to report when there is a reasonable probability that the use of, or exposure to, an article of food will cause serious adverse health consequences or death to humans or animals. Federal, state, and local government officials may voluntarily use the RFR portal to report information that may come to them about reportable foods.

FDA has received over 2240 RFR reports and has established new mechanisms for expedited evaluation and response so that swift action can be taken to protect consumers, typically in collaboration with the reporting firm. For example, RFR reports enabled FDA to act swiftly to address the presence of *Salmonella* in hydrolyzed vegetable protein, prepared foods containing undeclared sulfites, and glass in animal feed.

In 2011, FDA issued the first annual report on the RFR entitled, "A New Approach to Targeting Inspection Resources and Identifying Patterns of Adulteration," covering results from September 2009 through September

2010. The report demonstrates that, in its first year, the RFR significantly strengthened the ability of the FDA to track patterns of food and feed adulteration and target inspection resources to identify adulterated food/feed and prevent foodborne illnesses.

3. Antimicrobial Resistance

Foodborne pathogens that infect humans have become increasingly resistant to antimicrobial drugs, posing significant healthcare challenges when foodborne illness in humans requires treatment with those drugs. In order to address this problem, FDA, CDC and USDA participate in the National Antimicrobial Resistance Monitoring System (NARMS), a national public health surveillance system that tracks antibiotic resistance in foodborne bacteria. The most recent joint executive report on NARMS was published by FDA, CDC and USDA/ARS in May 2011. This report focuses on pathogens from food animals, retail meats and human clinical cases, based on data through 2008, and will provide a basis to inform and create scientifically sound solutions to public health and food safety concerns involving antimicrobial resistance.

The bacteria currently under surveillance by NARMS include: *Salmonella, Campylobacter, Escherichia coli,* and *Enterococcus.* NARMS monitors trends in antimicrobial resistance among foodborne bacteria from humans, retail meats, and animals; disseminates timely information on antimicrobial resistance to promote interventions that reduce resistance among foodborne bacteria; conducts research to better understand the emergence, persistence, and spread of antimicrobial resistance; and assists in making decisions related to the approval of safe and effective antimicrobial drugs for animals.

In addition to monitoring antimicrobial susceptibility, NARMS partners collaborate on epidemiologic and microbiologic research studies and examine foodborne bacteria for genetic relatedness, using pulsed-field gel electrophoresis (PFGE). NARMS also collaborates with antimicrobial resistance monitoring systems in other countries to work towards international harmonization of testing and reporting.

4. Import Safety

Food imports are growing rapidly and now comprise 15 percent of the total U.S. food supply, with much higher portions being imported in key categories, including: seafood (75-80 percent), fresh fruit (about 50 percent) and vegetables (about 20 percent).

FDA, FSIS, CDC, and U.S. Customs and Border Protection (CBP) have made significant efforts to improve U.S. oversight and ultimately the safety of imported food, including better targeting and data sharing at ports of entry, enhanced collaboration with foreign regulatory counterparts, and food safety capacity building where it is most needed in countries exporting to the United States.

In October 2010, several agencies entered into the Commercial Targeting and Analysis Center (CTAC) to expand access to import data. In addition, an Interagency Policy Coordination (IPC) committee was established by agency leadership as a part of the signed agreement on "Principles of Import." Three priorities have been identified by the IPC, including the Sea/Rail Manifest (document imaging), Cargo Control and Release (interface with PHIS), and the National Export Initiative of March 2010. Also, the U.S. Customs and Border Protection (CBP) has identified aggressive project schedules for the interim solutions over the next few months, which will enable shipments to be held at manifest level, receive documents through portal access to ACE, and interface with CBP's current IT system to enable data exchange when the entry is filed. In addition to the activities of this IPC, the FSIS will continue active involvement in the ACE/ITDS interim solutions.

FDA has developed and made significant progress in implementing a computer-based, data-driven tool for targeting examinations of imported food shipments, called PREDICT (Predictive, Risk-based Evaluation for Dynamic Import Compliance Targeting). This tool enables FDA to screen the over 10 million import entries of food commodities entering the U.S. annually to identify the ones most likely to pose a food safety risk or be in non-compliance with U.S. standards. PREDICT enables FDA to better protect food safety and make best use of its resources.

The food safety agencies have also strengthened international capacity and fostered cooperation with international partners to reduce global threats. The FDA continues to work closely with other countries, with a particular focus on those that export large volumes of foods to the United States. FDA also established a Joint Committee on Food Safety with its partner agencies in Canada (CFIA and Health Canada) to develop strategies to ensure the safety of our respective food supplies. FDA also continues to work closely with the appropriate entities in Mexico, including frequent communications on produce safety. The U.S. Deputy Commissioner for Foods visited China, the European Union, Canada and Mexico and continues to work to strengthen ties with these trading partners. The FDA foreign offices also foster critical communication and coordination on food safety efforts in China, India, Latin America, the

Middle East and Africa. In 2009, the CDC began providing training to epidemiologists in Central America to improve their ability to estimate the burden of foodborne illness, improve *Salmonella* surveillance and improve laboratory capacity in order to improve their own food supply, and reduce threats to the safety of foods which are likely to enter the global market. In April 2011, FSIS participated in a Central American Customs Union (CACU) workshop in El Salvador presenting information on product registration and export certification, as well as microbiological standards for meat products with particular emphasis on *Salmonella* and other pathogens. The regional workshop engaged the CACU countries of Costa Rica, El Salvador, Guatemala, Honduras and Nicaragua in crucial discussions of its own 2009 food safety standards. Their regulation seeks to establish the microbiological parameters and permissible limits for the registration and sanitary surveillance of food products and its provisions apply to all food products marketed in Central American countries.

CDC and FDA continue to be involved with and supportive of the important work of Global Foodborne Infection Network (formerly Global Salm Surv), INFOSAN, FERG, and other key World Health Organization (WHO) activities that support global food safety. For instance, in 2011, WHO Global Foodborne Infection Network held seven training courses in six countries. Each course brings together microbiologists and epidemiologists to learn methods for outbreak surveillance, and isolation, identification, and serotyping of various foodborne pathogens. In June 2009, the FDA established an agreement with the WHO for a global information platform and continues to develop a broad cooperative agreement between FDA and the WHO to coordinate work more efficiently.

The FDA has also developed the International Comparability Assessment Tool (ICAT) and a process for evaluating a foreign country's overall food safety regulatory system. From May through September of 2010, the ICAT and comparability process was piloted with New Zealand, which included extensive document review and on-site verification audits. A similar exercise began with the European Union in early 2011.

In December 2009, the FDA ran a third-party audit pilot program in shrimp processing facilities, many of which operate overseas. That program gave the FDA an opportunity to gather technical and operational information on facility compliance from third party auditors in order to develop evaluation tools and assess the utility of that information. The program also helped FDA assess the infrastructure needs for recognizing and managing third-party certification systems, gain knowledge of types of certification programs currently used by industry, and assess the implementation of certification programs against FDA's needs and expectations.

To further improve food safety practices in countries exporting to the United States, FDA, FSIS, the USDA Foreign Agricultural Service, and U.S. Codex office have created the International Policy Coordination Group (IPCG). The IPCG assesses ongoing U.S. Government (USG) international food safety technical assistance and capacity building activities, to develop a coordinated USG approach to developing and implementing future activities, and to consider future coordinated approaches to selected food safety issues facing developing countries and emerging markets. The IPCG is working on the transition from a forum for reacting to regulatory non-compliance to one of providing technical assistance focused on developing food safety systems, the operational principles for successful technical assistance, and current/future models to deliver effective programs. In addition, the IPCG is supporting the establishment of a unified International Food Safety Forum to help U.S. leadership work effectively with foreign counterparts to improve food safety internationally.

C. Food Safety Response

While much can be done to improve prevention, the highly complex and decentralized nature of the food supply system makes some occurrence of foodborne illness inevitable and impossible to eradicate completely. A sound food safety system depends on the quick and efficacious response to an outbreak once detected. Here, too, the FSWG agencies have taken important steps to increase the system's ability to respond to outbreaks of foodborne illness.

1. Strengthening the National Traceback and Response System

Creating a Coordinated Incident Command System

To facilitate communication and decision-making during outbreaks of foodborne illness, in 2010, HHS and USDA established an incident Command System Working Group" that developed protocols for a Multi-Agency Coordination Group for Foodborne Illness Outbreaks. This Coordination Group can quickly convene during an outbreak of foodborne illness involving multiple federal agencies to share information, make decisions, and leverage resources. In addition, outbreak response managers from CDC, FDA, and FSIS have been actively working together over the past two years to improve communications and coordination during outbreaks. FSIS and FDA have embedded epidemiologists in CDC's foodborne outbreak detection and response section to improve information flow among the agencies, especially during the early stages of an outbreak.

FDA has invested in a multi-tiered approach for the development and implementation of an effective NIMS Incident Command System. In order to promote the application of ICS principles within the agency and its federal, state, and local agency partners, FDA has delivered 27 classroom trainings of intermediate ICS 300 with a total of 753 personnel trained (694 FDA, 58 State personnel and one DHS) and 23 classroom training of advanced ICS 400 with a total of 592 personnel trained (571 FDA, 56 State personnel and one DHS), since mid-FY2008. Since March 2010, FDA has completed 8 sessions of ICS 402, providing an ICS overview for 109 Executives/Senior Officials within the agency. FDA has also delivered classroom position-specific training to a total of 105 personnel, including Incident Commanders, Operations Section Chiefs, Planning Section Chiefs, Logistics Section Chiefs and Finance/Admin Section Chiefs who are targeted to convene Incident Management Teams (IMTs) that will be expected to mobilize/deploy in a response capacity.

Updating Emergency Operations Procedures and Outbreak Response

An effective outbreak response requires quicker and better communication and coordination among federal, state, and local agencies. Since July 2009, CDC has expanded its partnership with state and local health departments to coordinate aggressive and rapid investigations of numerous major multistate outbreaks. These investigations have identified major hazards that required large recalls, such as the outbreak of *Salmonella* infections in 2010 that led to the recall of half a billion shell eggs, unexpected food vehicles, such as frozen microwaveable dinners, mamey fruit pulp, and Gouda cheese, and a new foodborne pathogen (Shiga-toxin producing *E. coli* O145, which was linked to shredded lettuce). In August 2009, the federal food safety agencies issued a letter to state and local agencies in support of the Council to Improve Foodborne Outbreak Response's (CIFOR) *Guidelines for Foodborne Disease Outbreak Response.* A CIFOR Guidelines Toolkit was published that will enable state and local jurisdictions to conduct a self-evaluation of their foodborne disease outbreak detection and response procedures and to identify recommendations in the CIFOR *Guidelines* that can improve their procedures. CDC funded 15 states and 4 local jurisdictions with small training grants that enabled their foodborne disease outbreak response team members to use the

Toolkit to self-assess and identify appropriate recommendations from the *Guidelines*. Due to the popularity of these *Toolkit* trainings, another RFP has been issued to fund additional state and local *Toolkit* trainings. The *Toolkit* has been distributed to all states and to all CIFOR member organizations and is available on the CIFOR website. In addition, FDA continues to work with State and local agencies to build food safety infrastructure nationally through the Integrated Federal-State Food Safety System effort.

FDA enhanced its role in outbreak response and prevention with the launch of the Coordinated Outbreak Response and Evaluation (CORE) Network. Staffed by experts in epidemiology, consumer complaints, statistics and veterinary medicine, the CORE Network is managing surveillance, response and post-response activities related to incidents of illness linked to FDA-regulated human and animal food. CORE's goals are to streamline decision-making, respond more quickly to outbreaks and ensure seamless coordination and enhanced communication. The team will also standardize post-response activities such as environmental assessments and root cause analyses, which provide an opportunity to learn what went wrong and use that information to drive strategies to prevent outbreaks from occurring in the future.

Since 2003, CDC has funded the National Environmental Health Association to conduct four Epi-Ready Foodborne Disease Outbreak Team Training courses per year for local and state environmental health specialists, laboratorians, and epidemiologists.

USDA has provided funding for live, interactive broadcasts of several of these Epi-Ready courses to remote sites that greatly increased the reach of the training. FDA staff have served as trainers as well as helping to guide the course content. The food safety agencies have also promoted a more highly trained environmental health workforce, which is skilled in properly conducting an environmental assessment during foodborne illness investigations. In total, more than 2,400 students from all 50 states have gone through Epi-Ready training.

In October 2009, the FDA expanded its federal-state Rapid Response Teams (RRT) project to help local and State partners work with FDA partners to identify and implement systems that strengthen existing state food safety programs and develop capabilities for coordinated and rapid responses to food-related incidents and emergencies. In FY 2010, the RRTs participated in an annual face-to-face meeting involving 80 participants from 40 federal and state offices. Meeting participants shared capabilities and spent time developing and harmonizing project directions. Since the meeting, the RRTs have formed working groups to develop a documentation of best practices in the development of key response capabilities. These documents include various measures that are being developed into metrics of capacity and achievement. Each working group completed a draft chapter for an "RRT Playbook" of rapid response capabilities, including the chapters on topics such as "Working with Other Agencies," "Communication Standard Operating Procedures (SOPs)," and "Joint Inspections."

Developing Industry Product Tracing Systems

Despite the dedicated efforts of food safety officials across the country, the existing capacity to traceback the sources of illness is limited. Public and private sector officials often lack rapid access to information about the sources of foods or ingredients, making the traceback process more cumbersome and leading to less-accurately targeted recalls. In addition, multiple federal, state, and local agencies all play essential roles in managing outbreaks, but lack a unified structure or adequate systems for sharing traceback data in an emergency. These limitations make it essential for Federal agencies to improve the traceability of food and the response to outbreaks of foodborne illness. In order to correct these problems, FDA has completed a pilot study on tracing with the tomato industry to devise more rapid and efficient ways to trace problems to their source. FDA and FSIS hosted public meetings on product tracing to identify steps the food industry can take to establish systems to improve our national capacity for detecting the origins of foodborne illnesses, and are currently reviewing comments received from the public meetings. In addition, FDA announced in September 2011 that the Institute of Food Technologists (IFT) will carry out two new pilot projects aimed at enhancing the agency's and industry's ability to trace products responsible for foodborne illness outbreaks. The pilots will evaluate methods and technologies for rapid and effective tracing of foods, including types of data that are useful for tracing, ways to connect the various points in the supply chain, and how quickly the data are made available to the FDA.

2. Recalls

FSIS revised the recall process to provide information to consumers more directly and clearly. The word "voluntary" was removed from recall communications to ensure consumers do not underestimate the seriousness of a recall. FDA and FSIS also collaborated on a mobile application for smart phones to provide consumers instant access to food recalls, alerts, and to view picture of labels. This application was launched by the White House in 2010, as a part of a product recall app, and can be accessed on USA.gov. In two recent outbreaks, inter-agency and industry collaboration provided consumers with a source to learn information about all of the recalled products so they could make informed food purchase and disposal decisions.

In order to broadly improve communication to the public regarding food safety, CDC, FDA and FSIS launched an enhanced and updated food safety website, "www.foodsafety.gov," in September 2009. The website also provides a mechanism for rapid information dissemination and alerts to consumers about food recalls.

D. Additional Agency Coordination, Capacity Building, and Partnerships

Partnerships among federal agencies, state, local, tribal, and territorial entities, academia, and private partners are crucial to enhancing the safety of the food supply.

Laboratory practices and capacity are key areas for inter-agency collaboration. FSIS, FDA and CDC have had ongoing technical exchanges to promote strategies for rapid detection of *Salmonella* in food products and emerging technologies for molecular sub-typing methods. These efforts are coordinated with industry to more effectively characterize pathogens during outbreak investigations. FSIS and the Food Emergency Response Network (FERN) laboratories have developed and established national standards for laboratory qualifications for detection of pathogens in food involving 37 federal agencies; 116 state and 17 local laboratories; and signed 30 cooperative laboratory agreements. FDA developed a Partnership for Food Protection (PFP) workgroup to develop national laboratory standards to address consistency of findings among state laboratories.

FSIS and FDA are also assisting state food testing laboratories in obtaining accreditation under ISO Standard 17025, the gold standard for food-testing laboratories. This is part of a multi-year effort to create additional opportunities to identify means to protect public health. In addition, the FERN is equipped to provide supplemental laboratory analytical and surge capacity during large-scale events, both natural and man-made. FERN has been successfully activated during several recent incidents including melamine, *E. coli* O157:H7 in spinach, and the oil spill. FDA has established "High Throughput Laboratories" that cover the areas of Chemistry, Microbiology, Drugs and Environmental Samples. These laboratories have established capabilities to handle larger volumes of samples in shorter time-frames without compromising quality or integrity. Development also continues on the FDA Laboratory Information System (LIMS) initiative, under the Automated Lab Management (ALM) Program, to provide an automated system to manage and automate laboratory operations throughout all 13 FDA field and 2 mobile laboratories.

In FY 2011, FDA established Vet-LRN, the Veterinary-Laboratory Response Network, which includes state and federal laboratories that integrate resources and expertise for timely and accurate reporting, identification, and analysis of animal feed chemical and microbiological contamination events. The system operates by examining animal tissues and diagnostic specimens for microbiological agents, toxins, and other causes of disease. The Center for Veterinary Medicine provides early detection of pet food borne disease outbreaks with rapid notification to stakeholders in order to minimize animal illness and related economic loss. These efforts contribute to overall food safety as animal feed events could signal potential issues in the human food system.

In March 2011, Vet-LRN held its first developmental meeting with veterinary laboratory directors from around the U.S. and Canada. The goal of the meeting was to establish contact with various laboratories that are interested in joining the network. Comments and ideas were provided by the laboratory directors to help Vet-LRN plan its activities and coordinate with other existing networks such as the Food Emergency Response Network and other animal disease health networks in the United States such as USDA's National Animal Health Laboratory Network.

CDC continues to strengthen PulseNet, the national molecular subtyping network for foodborne disease surveillance which has revolutionalized the detection and investigation of foodborne disease outbreaks. This network of local, state, territorial, agricultural and federal laboratories in 50 states and 82 countries is coordinated by CDC and the Association of Public Health Laboratories.

Beyond laboratory collaborations, several forums have been created or reconfigured to better enable agency coordination. For instance, CDC, FDA and FSIS convened an interagency outbreak response working group to clarify roles and interactions among agencies during outbreak response activities. This effort led to earlier and routine communications among agencies in outbreak detection, and a program to provide cross-training of staff among agencies.

The Interagency Risk Assessment Consortium (IRAC) has been restructured to serve as the coordinating council to develop and oversee the conduct of joint FDA-FSIS-CDC risk assessments. The IRAC includes membership from 19 federal agencies and sub-agencies and meets quarterly to review progress in accomplishing annual goals on specific topic areas and activities relevant to food-safety risk assessments. This includes developing risk assessment modeling methodology, engaging in data collection and analysis, and providing interagency peer review of risk assessments.

Another forum, the Food Safety Working Group Information Technology Task Force (ITTF), was established to develop recommendations for achieving greater interoperability and harmonizing electronic data collection standards between the agencies and State and local authorities. Utilizing a useful crowd-sourcing tool, IdeaJam, the ITTF developed 41 initial recommendations, thereby fulfilling its charge and eventually agreeing to implement 15 final recommendations approved by the CDC, EPA, FSIS, and FDA.

In August 2010, FDA hosted a 50-state workshop entitled "A United Approach to Public Health" to facilitate greater progress toward the creation of an integrated national food safety system. Building upon the 2008 workshop, where ten working groups were formed under the Partnership for Food Protection (PFP), this workshop highlighted the progress of the PFP workgroups. These workgroups consisted of Response, Training, Interactive Information Technology, and Risk-Based Work Planning; and the FDA/ Center for Veterinary Medicine-led Pet Event Tracking Network (PETNet) project. The Partnership for Food Protection Coordinating Committee is currently looking at the recommendations made by the 50-state breakout groups and is evaluating how best to integrate the suggestions to improve the design and implementation of an Integrated Food Safety System.

E. Retail Food Safety

In September 2011, FDA announced a Retail Food Safety Action Plan that includes several measures to help assure the safety of food sold in food stores, restaurants, schools, and other foodservice operations in the United States. The Action Plan focuses on improving the way managers of these establishments conduct food safety operations in their facilities, as well as improving the oversight of these establishments by public health agencies at the federal, state and local levels. The Plan specifically calls for strengthening State and local food safety requirements that apply to these establishments and for improving training for personnel on measures to keep food safe.

F. Consumer Education

While those involved in the commercial production, processing, distribution and retail sale of food have a primary obligation to do everything they reasonably can to prevent food safety problems, consumers also contribute to prevention by observing proper food handling practices. The federal government is enhancing its support for food safety education that enables consumers to better understand and play their food safety role.

To further engage consumers and educate them about the importance of food safety in the kitchen, USDA, in collaboration with FDA and CDC, launched an Ad Council advertising campaign to educate consumers about safe food handling practices. This Food Safe Families campaign is a national, multi-media effort utilizing television, radio, print and online advertising designed to fulfill the agencies' responsibility to raise awareness of foodborne illness and provide consumers with information they need to practice safe food handling behaviors. The ads urge consumers to learn more at http://www.foodsafety.gov—a redesigned and vastly expanded inter-agency website which hosts vital food safety information on safe handling behaviors and recall information. This unprecedented campaign exemplifies inter-agency cooperation. The campaign also benefited from input of the Ad Council's Expert Panel which included a vast array of stakeholders all along the farm to table continuum including producers, industry and consumer groups, retailers, academics and public health professionals.

The FDA awarded the Partnership for Food Safety Education (PFSE) a sole-source grant to support a strategic planning process to determine future directions and strategies for food safety education. Through the strategic planning process, technical experts have examined the scientific underpinnings of current messages, while social scientists have examined strategies for creating behavior change. PFSE is a not-for-profit organization that unites industry associations, professional societies, consumer groups, academia and government to educate the public about safe food handling.

FDA also conducted its 12th annual, week-long training session for science teachers based on its *Science and Our Food Supply* curriculum. Science teachers learn the fundamental science critical to food safety and perform the curriculum laboratories. They return to their schools around the country and teach students important safe food handling measures to prevent illness and educate them on the science of food safety. The training was developed in cooperation with the National Science Teachers Association. Since the program began, the curriculum has reached about 4 million students nationwide. In addition, FSIS and FDA updated a series of new brochures for populations who have been identified as having at greater risk of foodborne illness and posted on its web site two videos for consumers on the safe handling of seafood and fresh produce.

FSIS conducted consumer focus-group studies in 2010 to measure consumer understanding of labeling regarding the safe handling and preparation of meat, poultry, and egg products. Participants provided useful information on their lack of understanding of labeling information. The findings of this focus group study can be used in the future to improve decisions about labeling and to better develop a consumer survey that could then be generalized to the consumer population.

CDC's new *Vital Signs* is a monthly report on a single, important public health topic. In June, 2011 the *Vital Signs* report focused on Food Safety. The report was led by CDC's Foodborne Diseases Active Surveillance Network (FoodNet) program and was a coordinated effort with FDA and USDA. The data highlights the success in reducing *E. coli* O157 infections, while pointing out that *Salmonella* infections have not declined in 15 years. The report takes consumers through the steps along the farm-to-table continuum and shows what needs to be targeted for action everywhere food is grown (production), made (manufactured), moved (transportation), prepared (in restaurants, grocery stores, and homes), and consumed. The report ends with an important Call to Action for everyone who is involved in food safety and points to FoodSafety.gov as a gateway to information for consumers, including a blog on *Salmonella.* The report was released in the midst of the *E.coli* O104 outbreak in Europe, calling even more attention to this important public health concern, and reached over 684 million people within 48 hours. (Source: http://www.foodsafety.gov/keep/basics)

II. Food Safety Working Group 2011-12 Agenda and Beyond

The FSWG's accomplishments to date represent a large down payment on a stronger food safety system that will deliver greater value, better prevent illnesses and more effectively promote the well-being of the American people. Building on those efforts, the FSWG will continue to strengthen the food safety system through increased prevention, enhanced surveillance, and faster response. It will do so in part through implementing the new FDA Food Safety Modernization Act (FSMA).

During 2010, the FSWG's member agencies worked closely with the Administration, Congress, and key stakeholders on this critical legislation. The FSMA provides a new vision and mandate for the food safety system, focused on science-based prevention of food safety problems and risk-based targeting of public and private prevention efforts to get the most risk reduction "bang for the buck."

To carry out this vision, Congress has directed FDA to establish new prevention oriented standards for the food industry, provided new inspection and enforcement tools to ensure high rates of compliance with those standards, and mandated the creation of a new system of import oversight so that imported food meets the same modern standards as domestically-produced food. FSMA also calls for FDA to work in close partnership with other federal agencies, to build a nationally integrated system of inspection and laboratory testing with state and local partners, and to work closely with foreign governments to improve import safety. In addition, Congress has directed CDC to strengthen and enhance public health surveillance and response systems in partnership with state and local health departments and to designate Centers of Excellence which will provide additional resources for frontline health professionals during outbreaks and conduct research and outreach activities regarding food safety.

The FSMA presents a significant opportunity and challenge for our food safety agencies. Implementation of this new legislation will be one of the highest priorities for the FSWG and its members over the next several years.

A. Greater Prevention

The federal food safety agencies will build on past prevention efforts and harness the new mandates of the FSMA to strengthen prevention of foodborne illness.

1. Pre-Harvest Food Safety

FSIS, FDA and CDC have a common interest in working with the scientific, agricultural and public health communities to solve the problem of infection and transmission of foodborne disease organisms at the point of livestock and fresh produce production. For example, while *E. coli* O157:H7 emerged as a hazard in beef in the early 1990s, the pathogen is now a significant cause of illness associated with fresh produce as well. While FSIS will continue implementing the FSWG recommendation to increase enforcement in beef facilities and FDA will issue rules for grower practices affecting produce safety, FSIS and its partner agencies at USDA will work with other federal agencies, producers, and scientists to discuss how to minimize pathogen contamination in animal production. FSIS and other USDA agencies will take the lead in conducting a thorough review and assessment of pre-harvest activities in the industry and government and engage stakeholders to consider initiatives in the areas of research, incentives for technology development and adoption, and identification and dissemination of best practices.

USDA convened a public meeting in November among FSIS, the Animal and Plant Health Inspection Service (APHIS), and the Agricultural Research Service (ARS) to discuss how pre-harvest pathogen control strategies for animals presented for slaughter can reduce the likelihood that beef could become contaminated with Shiga toxin-producing E. coli, Salmonella, and other pathogens. The meeting featured presentations on the latest research, followed by workshop discussions that included a wide variety of stakeholder groups. After analyzing the public comments presented at the meeting, USDA intends to convene a second public meeting focused on pre-harvest pathogen control strategies for poultry.

2. Upcoming Preventive Control Standards

As mandated by the FSMA, FDA will issue new rules establishing preventive control requirements for produce growers, food and animal feed processing facilities, and food transporters, as well as to prevent intentional adulteration of food over the next one to three years. FDA has done significant outreach with stakeholders to prepare for the issuance of these rules. It held a series of three public meetings to inform interested parties of the agency's current thinking regarding preventive controls, import oversight, and inspection and compliance and to solicit comment from stakeholders. The public meeting on preventive controls, in particular, gave stakeholders an opportunity to hear about FDA's thinking on preventive controls for food and feed facilities and produce safety standards. This public meeting complemented the listening tours with farmers that FDA undertook in 2010 and 2011 in thirteen states to understand the complexities of the produce industry prior to developing a regulation. FDA is working towards a release of proposed regulations on preventive controls for food and feed facilities, produce safety standards, and a foreign supplier verification program to ensure that importers are verifying compliance with these standards for imported foods. FDA will work with USDA and other Administration partners to ensure that the final standards take into account the full diversity and complexity of these sectors.

FDA has also recently established a National Food Safety Preventive Controls Alliance and an industry-oriented food safety training program to assist the US food industry in complying with regulations promulgated in response to FSMA legislation. The objectives of the National Food Safety Preventive Control Alliance include:

- To provide food facilities in the US with the resources to be in compliance with the preventive control component of FSMA.
- To assist the FDA to disseminate the science and technical elements relevant to the hazard analysis and preventive controls aspects of FSMA legislation to the US food industry.
- To assist the US food industry, particularly the small and medium-sized companies, to be compliant with the FSMA legislation.

For egg safety, FSIS is developing a proposed rule to address food safety risks in the egg products industry and will require Hazard Analysis and Critical Control Point (HACCP) systems in every establishment that produces egg products. By applying new standards, the egg product industry will be expected to comply with a system similar to that for meat and poultry products. FDA will continue to implement its prevention-based regulations for shell eggs.

3. Retail Food Safety

The federal agencies participating in FSWG have several proposed rules and other initiatives underway that will have a significant impact on retail food safety. For instance, within the coming year, FDA will improve retail food safety by encouraging more uniform state adoption of FDA's recommended standards for retail food safety, strengthening state and local inspection programs, and increasing the presence of certified food safety managers in retail facilities in accordance with its Retail Food Safety Action Plan. FSIS will publish a new regulation to revise its inspection procedures to ensure they are better focused on public health protection. FDA, FSIS and CDC are working on a risk assessment on *Listeria monocytogenes* that specifically addresses retail practices. In preparation for this risk assessment, a *Federal Register* notice requesting scientific data was published and the agencies held a public meeting. In addition the agencies are working through CDC's Environmental Health Specialists Network (EHS-Net) program and engaging in collaborative research with industry and academic partners to develop supporting data.

B. Enhanced Surveillance and Compliance

1. Domestic Inspection and Compliance

Under the FSMA, FDA will be modernizing its approach to food safety inspection to take advantage of the preventive control framework and make better use of scarce inspection resources. The primary focus will shift from looking for problems in food safety facilities and correcting them after the fact to verifying that facility operators are implementing well-planned and documented systems for preventing problems. This approach is more effective and efficient because it helps ensure that proper practices are being observed on a continuing basis.

FDA will also begin using the new administrative enforcement powers granted by the FSMA to prevent food safety problems. This includes administratively detaining products that have been produced under sub-standard conditions that jeopardize safety and suspending the registration and thus ability to operate facilities whose food products are putting consumers at risk due to inappropriate practices. When necessary, FDA will also use its new power to mandate recalls of foods that are contaminated or linked with illness outbreaks.

2. Import Safety

FDA will focus on implementing the new import safety tool kit Congress created with the enactment of FSMA. Under the new system, food importers will be responsible for providing documented assurances to FDA that the food they import has been produced under the same prevention-oriented standards as domestic food. FDA will be able to verify the adequacy of the assurances by examining the importer's records and selectively examining import shipments. FDA will also support and supplement the efforts of importers by establishing an accredited third-party certification program, working with foreign governments and assessing the adequacy of food safety oversight and practices in countries exporting to the United Sates, and conducting inspections of foreign food facilities. Finally, FDA will implement a system to expedite entry of food shipments for importers that have especially well-documented systems to ensure safety.

FSIS has been working to further define its risk-based methodology for audits of equivalent countries allowed to export product to the U.S. FSIS will publish a document that details a performance-based approach to audits of foreign countries and point-of-entry re-inspections. This documentation will ensure that foreign equivalence audits continue to move in the direction of a risk-based approach and focus resources in an effective and efficient manner.

Federal agencies will continue working to ensure that foreign governments have the technical expertise and understanding of U.S. requirements that they need to be effective food safety partners.

3. Foodborne Illness Surveillance and Incident Investigation

As funding becomes available, CDC will maintain, upgrade and expand the PulseNet, and other subtyping networks, with more participants and next-generation methods to make outbreak detection and investigation faster and more robust for more pathogens. For example, CDC is developing new methods to deploy in state health department laboratories that will rapidly identify and subtype the important non-O157 Shiga toxin-producing *E. coli*, such as the one that caused a large and deadly outbreak in Germany in 2011. FoodCORE will evaluate the impact of enhanced outbreak response activities so that the most successful methods can be adopted by other state and local health departments. The proportion of foodborne illness that can be attributed to specific food commodities will be estimated based on reported foodborne outbreaks. In addition, CDC will launch a new FoodNet case-control study of risk factors for infection with non-O157 Shiga toxin producing *E. coli* and will complete a study of risk factors for developing hemolytic uremic syndrome among persons with *E. coli* O157:H7 infections.

In order to improve prevention and surveillance efforts, food safety programs need additional and new information on contributing factors and environmental antecedents of foodborne illness outbreaks. Currently, this information is lacking. A national voluntary environmental assessment information system could provide food-safety program managers with an information resource that could fill this gap. As a way to foster wider use of environmental assessments of farms and facilities to identify possible pathways of contamination, the National Voluntary Environmental Assessment Information System (NVEAIS) will be launched next year. CDC, as the lead agency, will continue to work with FSIS, FDA, state EHS-Net sites, and membership of the Conference for Food Protection to develop and implement the system. Information collected through NVEAIS will be used to establish a detailed characterization of food vehicles and monitor food vehicle trends, identify and monitor contributing factors and their environmental antecedents, and provide a basis for hypothesis generation regarding factors that may contribute to foodborne outbreak events. With this information, food safety programs and the food industry will have data to guide the planning, implementation and evaluation of foodborne illness prevention activities.

With support from FDA, FSIS, the National Park Service and EHS-Net states, CDC has also developed a virtual-world training program on how to conduct foodborne illness outbreak environmental assessments. This training program will be one of several requirements for participation in NVEAIS.

4. Product Tracing

FSIS relies heavily on records maintained by industry to identify trace back and trace forward on FSISregulated products associated with foodborne illness and other food safety incidents. Retail records are a critical component in trace back and trace forward activities, and are essential to quickly and effectively determine source product and ensure controls are enhanced by affected product manufacturers (e.g., official establishment, retail, foodservice). Yet, recent outbreak investigations were impeded by poor retail records.' FSIS will propose a rule to enhance access to records to facilitate trace back in case an illness or outbreak is associated with ground beef from retail stores. FSIS will also develop compliance guidelines that retailers can use to meet FSIS trace back and trace forward activities, and additional guidance for investigators focused on activities at the retail level.

As mandated by the FSMA, FDA will consider information gathered through pilot tests of approaches to effective product tracing in other food categories, work with the food industry to foster innovative approaches to improve tracing, and improve its internal systems for tracing food products to their origin.

C. Improved Response

1. Outbreak Response

By all accounts, the government response to recent multistate outbreaks has been effective in swiftly reducing risks and saving lives. Our new systems were put to the test with the recent Listeria outbreak in cantaloupes. Enhanced surveillance, coordinated by CDC, and a rapid, multistate response were critical in controlling the outbreak. Close collaboration between FDA, CDC and the state of Colorado resulted in quick identification and recall of the contaminated cantaloupes of the source of the contaminated cantaloupes, enabling consumers to be advised quickly to avoid the contaminated food, and helping to prevent further exposures, illnesses, and deaths.

Within a week of FDA issuing a press release announcing the single source of contaminated cantaloupes and that the cantaloupes were recalled, FDA had a multi-disciplinary team on the ground conducting an environmental assessment to better understand how the contamination occurred. Updates on the recall and sub-recalls are posted on FDA's website to ensure consumer awareness.

We continue to dedicate resources and expertise to learn all that we can from these outbreaks, in order to identify risks in advance of future outbreaks. FDA evaluates information, including environmental and product samples used to determine the root-cause of how whole cantaloupe became contaminated with Listeria. It is clear that we have already realized positive gains from our investment to develop an improved system. For instance, the funding we provided to the state of Colorado to improve surveillance for foodborne illnesses through its participation in the FoodNet active surveillance system, a program sponsored by CDC, FDA, and USDA was instrumental in enabling officials in Colorado to quickly recognize the outbreak of listeriosis cases and conduct interviews to help identify the most likely food associated with the illnesses.

FDA has also recruited a Chief Medical Office/Director of Outbreaks, who will lead a new Coordinated Outbreak Response and Evaluation (CORE)

Network within FDA to ensure rapid and effective emergency response and more systematic follow up investigations, in collaboration with CDC and other agencies. Future prevention efforts will be created based on lessons from past outbreak experiences. CDC will continue to evaluate the best methods in FoodCORE Sentinel Sites and will promote best practices among all local and state health departments.

2. Data Analysis

In April 2011, FSIS launched the Public Health Information System (PHIS) to help respond more rapidly to current and potential food safety threats. PHIS, which will strengthen FSIS detection and response to foodborne hazards, will be a flexible, user-friendly, and web-based application that replaces many of FSIS' legacy systems, such as Performance Based Inspection System (PBIS) and the Automated Import Information System (AIIS), automates paper-based business processes, and can be modified to accommodate changing needs. PHIS uses a systems approach to food safety. Through its predictive analytics component, PHIS will integrate FSIS' data streams. This function will support a data-driven approach to FSIS inspection, auditing, and scheduling and result in a comprehensive, timely, and reliable data-driven inspection system.

Once fully implemented, PHIS will revolutionize the agency's ability to utilize data in real time to inform all aspects of its domestic inspection, import inspection, and export activities.

This system will make the Agency and its employees more accountable, and allow FSIS to collect more information about the U.S. domestic and international food safety systems, which produce FSISregulated products. Using multiple data sources, PHIS will allow analysts to identify trends that will provide the agency with the capability to adjust domestic and import inspection and sampling. As a result, FSIS will be in a position to better identify food safety risks and detect problems before they reach consumers and result in outbreaks and recalls. PHIS will enable alerts for imported product to be triggered by real time data monitoring and will automate FSIS' risk-based approach to foreign country audits and re-inspection. In addition, work will continue under the Data Analysis and Integration Group (DAIG) at FSIS to coordinate the agency's data collection, analysis, and integration activities across all program areas. Also, the FSIS, FDA and CDC have established the Interagency Food Safety Analytics Collaboration (IFSAC) group to develop and share analytical methods, common terminology, and standards of practice. The IFSAC will address key issues such as foodborne illness attribution.

D. Consumer Education

Consumers play a key role in the farm-to-table approach to food safety. They are the last step where food can be contaminated to a level that could cause harm and where proper handling can minimize the risk of harm. Consumer education will therefore continue to be an important part of prevention, as well as the target for information during a food recall or outbreak to identify and prevent use of products that should not be consumed. The FSWG agencies will focus on three areas for improving consumer education:

- elevating interagency and community-based partnerships for food safety education;
- expanding the science base and reach of consumer education; and
- strengthening risk communication related to foodborne outbreaks and recalls.

E. Partnerships

A central element of the vision laid out by the FSMA is that no one agency, level of government or private sector initiative can succeed in meeting today's food safety challenges alone—collaborative action and effective partnerships are vital. Food safety presents many challenges for the entire food system, public and private, and from farm-to-table. Meeting the challenges across organizational lines within government and between government and the private sector is the vision of the FSWG, and the members of the FSWG are committed to fulfilling it.

Many ongoing partnership efforts are outlined in this report. Future efforts will involve the many instances in which FSMA mandates inter-agency collaboration and coordination among federal agencies on such topics as produce safety, preventive controls in food facilities, intentional adulteration standards, technical assistance for small growers and facility operators, and improving surveillance of foodborne illness.

The partnership efforts will also focus heavily on strengthening the capacity of state and local agencies and integrating federal, state and local efforts for a more effective and efficient food safety system, as well as collaborating with foreign governments of import safety.

Finally, government must continue to partner with the food industry and consumers to ensure food safety. Industry bears the primary burden and responsibility to produce safe food, and has vast experience and expertise in doing so. Consumers are also critical partners in the food safety system and in ensuring better outcomes. To be successful, government must further enlist the effort, expertise and perspectives of all stakeholders through systematic outreach and active listening. The FSWG and its members are committed to fostering and maintaining that partnership to sustain and enhance our current system, and continue to provide one of the safest food supplies in the world.

End Note

[1] To illustrate, FSIS's Office of Public Health Science investigated 16 cases of foodborne illness implicating raw ground beef products manufactured at retail in 2007-2008. Of the 16, only 9 retail operations kept production logs (e.g., grinding logs) sufficient for trace back and trace forward activities. Of the 9, 5 resulted in a recall. (See www.fsis.usda.gov/PPT/ Recordkeeping_Presentation.ppt).

In: The Federal Food Safety Working Group ISBN: 978-1-62417-059-1
Editors: R. Jaynes and S. Darlington

Chapter 2

Federal Food Safety Oversight: Food Safety Working Group Is a Positive First Step but Governmentwide Planning Is Needed to Address Fragmentation*

United States Government Accountability Office

Why GAO Did This Study

For more than a decade, GAO has reported on the fragmented nature of federal food safety oversight and how it results in inconsistent oversight, ineffective coordination, and inefficient use of resources. In 2007, GAO added this issue to its high-risk list. In March 2009, the President established the Food Safety Working Group (FSWG) to coordinate federal efforts and establish food safety goals to make food safer.

Section 21 of Public Law 111-139 mandated that GAO identify programs, agencies, offices, and initiatives with duplicative goals and activities. This review examines: (1) steps, if any, that the FSWG has taken to increase collaboration among federal food safety agencies, and (2) options we and others have identified to reduce fragmentation, overlap, and potential

* This is an edited, reformatted and augmented version of United States Government Accountability Office, Publication No. GAO-11-289, dated March 2011.

duplication in food safety oversight. GAO reviewed information about the FSWG and alternative organizational structures for food safety, and conducted interviews.

WHAT GAO RECOMMENDS

GAO recommends that the Director of OMB, in consultation with the federal food safety agencies, develop a governmentwide performance plan for food safety that includes results oriented goals and performance measures for food safety oversight and a discussion about strategies and resources. OMB declined to comment on a draft of this report. USDA and Health and Human Services provided technical comments.

WHAT GAO FOUND

Creation of the FSWG elevated food safety as a national priority, demonstrated strong commitment and top leadership support, and was designed to foster interagency collaboration on this cross-cutting issue. The FSWG includes officials from the Food and Drug Administration (FDA), the U.S. Department of Agriculture (USDA), the Office of Management and Budget (OMB), and other federal agencies. Through the FSWG, federal agencies have taken steps designed to increase collaboration in some areas that cross regulatory jurisdictions—in particular, improving produce safety, reducing *Salmonella* contamination, and developing food safety performance measures.

However, the FSWG has not developed a governmentwide performance plan for food safety that provides a comprehensive picture of the federal government's food safety efforts. When GAO added food safety oversight to its high-risk list in 2007, it said that what remains to be done is to develop a governmentwide performance plan for food safety that is mission based, results oriented, and provides a cross-agency perspective. Officials from OMB, FDA, and USDA told us that the FSWG's July 2009 "key findings" represent the governmentwide plan for food safety. However, most of the goals outlined in the key findings are not results oriented and do not include performance measures. Further, the FSWG has not provided information about the resources that are needed to achieve its goals. Our prior work has identified results oriented goals and performance measures and a discussion of strategies and resources as standard elements of performance plans.

GAO and other organizations have identified options to reduce fragmentation and overlap in food safety oversight in the form of alternative organizational structures, but a detailed analysis of their advantages, disadvantages, and potential implementation challenges has yet to be conducted. GAO has suggested that Congress consider commissioning the National Academy of Sciences or a blue ribbon panel to conduct a detailed analysis of alternative organizational structures for food safety. Some of the alternative organizational structures include a single food safety agency, a food safety inspection agency, a data collection and risk analysis center, and a coordination mechanism led by a central chair. GAO recognizes that reorganizing federal food safety responsibilities would be a complex process that could have short-term disruptions and transition costs. GAO and other organizations have regularly paired proposals for alternative food safety organizations with calls for comprehensive, unified, risk-based food safety legislation. New food safety legislation that was signed into law in January 2011 strengthens a major part of the food safety system; however, it does not apply to the federal food safety system as a whole or create a new risk-based food safety structure.

Abbreviations

AMS	Agricultural Marketing Service
CDC	Centers for Disease Control and Prevention
DOD	Department of Defense
EU	European Union
FDA	Food and Drug Administration
FSIS	Food Safety and Inspection Service
FSWG	Food Safety Working Group
OMB	Office of Management and Budget
USDA	Department of Agriculture
VA	Department of Veterans Affairs

Congressional Committees

For more than a decade, GAO has reported on the fragmented nature of federal food safety oversight. While the Food and Drug Administration (FDA) and U.S. Department of Agriculture's (USDA) Food Safety and Inspection Service (FSIS) have primary oversight responsibilities, a total of 15 agencies collectively administer at least 30 food-related laws. In 2007, we added the federal oversight of food safety to our list of high-risk areas in need of broad-based transformation to achieve greater economy, efficiency, effectiveness, accountability, and sustainability.[1] We also cited the need to integrate this fragmented system as a challenge for the 21st century.[2]

The 2010 nationwide recall of more than 500 million eggs due to *Salmonella* contamination highlights how this fragmentation persists, with several agencies having different roles and responsibilities throughout the egg production system. For example, FDA is generally responsible for ensuring that eggs in their shells—referred to as shell eggs—including eggs at farms such as those where the outbreak occurred, are safe, wholesome, and properly labeled. FSIS, on the other hand, is responsible for the safety of eggs processed into egg products. In addition, USDA's Agricultural Marketing Service (AMS) sets quality and grade standards for shell eggs, such as Grade A, but does not test the eggs for bacteria such as *Salmonella*. Further, while USDA's Animal and Plant Health Inspection Service manages the program that helps ensure laying hens are free from *Salmonella* at birth, FDA oversees the safety of the feed they eat.

Many of our reports have found that fragmentation in the nation's food safety system results in inconsistent oversight, ineffective coordination, and inefficient use of resources. In 2005, we reported that federal agencies spend resources on overlapping activities, including inspections of domestic and imported foods, training, research, risk assessment, education, and rulemaking.[3] We recommended that federal agencies take a number of steps to reduce overlap, leverage resources, and enhance coordination. For example, to better use FDA's limited inspection resources and leverage USDA's resources, we recommended that, if appropriate and cost effective, the Commissioner of the FDA enter into an agreement to commission USDA inspectors to carry out FDA's inspection responsibilities for food establishments that are under the jurisdiction of both agencies. We have made several other recommendations intended to address the fragmented federal oversight of the nation's food supply. In 2001, we recommended that the Secretary of Agriculture, the Secretary of Health and Human Services, and the Assistant to the President for Science and Technology, as joint chairs, reconvene the President's Council on Food Safety to facilitate interagency coordination on food safety regulation and programs.[4] The council had disbanded earlier that year. Positively, in March 2009, the President established the Food Safety Working Group

(FSWG), which is co-chaired by the Secretaries of Agriculture and Health and Human Services, to coordinate federal efforts and develop goals to make food safer. When we added food safety oversight to our high-risk list in 2007, we said that what remains to be done is to develop a governmentwide performance plan for food safety that is mission based, has a results orientation, and provides a cross-agency perspective.[5] Finally, in reports dating back to 2001 we have been suggesting that Congress consider commissioning the National Academy of Sciences or a blue ribbon panel to conduct a detailed analysis of alternative organizational food safety structures and consider enacting comprehensive food safety legislation.[6]

New food safety legislation that was signed into law in January 2011—FDA Food Safety Modernization Act—strengthens a major part of the food safety system.[7] It shifts the focus of FDA regulators from responding to contamination to preventing it, according to FDA, and expands FDA's oversight authority. For example, the law directs FDA to increase the frequency of its inspections, and allocate resources to inspect facilities according to the facilities' known safety risks, with high-risk facilities being inspected the most frequently. The law also has several sections that require interagency collaboration on food safety oversight in areas such as inspections, seafood safety, and food imports. While the new law is a positive development, it does not apply to the federal food safety system as a whole. In particular, it does not address USDA's authorities, which remain separate and distinct from FDA's.

A new statutory requirement mandates that GAO identify federal programs, agencies, offices, and initiatives with duplicative goals and activities within departments and governmentwide.[8] Under that mandate, this review examines: (1) steps, if any, that the FSWG has taken to increase collaboration among federal food safety agencies, and (2) options we and others have identified to reduce fragmentation, overlap, and potential duplication in food safety oversight. To complete our work we reviewed food safety reports and legislation, and interviewed officials from USDA, FDA, and the Office of Management and Budget (OMB). We also collected and analyzed information about the FSWG, its activities, and its governmentwide plan for food safety, as well as alternative organizational structures for food safety. More detailed information about our scope and methodology appears in appendix I. We conducted this performance audit from July 2010 to March 2011 in accordance with generally accepted government auditing standards. Those standards require that we plan and perform the audit to obtain sufficient, appropriate evidence to provide a reasonable basis for our findings and conclusions based on our audit objectives. We believe that the evidence obtained provides a reasonable basis for our findings and conclusions based on our audit objectives.

BACKGROUND

The safety and quality of the U.S. food supply is governed by a highly complex system stemming from at least 30 laws related to food safety that are collectively administered by 15 agencies. The two primary food safety agencies are USDA and FDA. USDA is responsible for the safety of meat, poultry, processed egg products, and, as soon as recently proposed regulations are finalized, catfish. FDA is responsible for virtually all other food, including seafood. In fiscal year 2009, budget obligations for FDA's Foods Program were $685 million and budget obligations for FSIS were approximately $975 million.[9] Federal food safety activities include inspecting domestic food-processing facilities and imported food at ports of entry, visiting foreign countries or firms to inspect and evaluate foreign food safety systems, analyzing samples collected at food-processing facilities to identify possible contamination, rulemaking and standard setting, and developing guidance for industry, among other things. Appendix II summarizes the federal agencies' food safety responsibilities and main authorizing statutes. The federal food safety system is supplemented by the states, which may have their own statutes, regulations, and agencies for regulating and inspecting the safety and quality of food products.

The existing food safety system, like many other federal programs and policies, evolved piecemeal, typically in response to particular health threats or economic crises. Existing statutes give agencies different regulatory and enforcement authorities for different food products. For example, the 2008 Farm Bill gave USDA responsibility for inspecting catfish, but left general responsibility for seafood safety with FDA, making the system more fragmented. According to USDA officials, USDA estimates it will spend no more than $5 million in fiscal year 2011 and did not request funding for fiscal year 2012 for its catfish inspection program.[10]

Three major trends also create food safety challenges. First, a substantial and increasing portion of the U.S. food supply is imported. Second, consumers are eating more raw and minimally processed foods. Third, segments of the population that are particularly susceptible to food-borne illnesses, such as older adults and immune-compromised individuals, are growing.

The FSWG Has Taken Steps Designed to Improve Collaboration among Federal Agencies but Has Not Developed a Comprehensive Governmentwide Performance Plan for Food Safety

Creation of the FSWG by the President in March 2009 elevated food safety as a national priority, demonstrated strong commitment and top leadership support, and was designed to foster interagency collaboration on this crosscutting issue. However, the FSWG has not developed a governmentwide performance plan that provides a comprehensive picture of the federal government's food safety efforts. The FSWG is co-chaired by the Secretaries of Health and Human Services and Agriculture. It also includes officials from FDA, FSIS, the Centers for Disease Control and Prevention (CDC), the Environmental Protection Agency, the Department of Homeland Security, the Department of Commerce, the Department of State, and several offices in the Executive Office of the President, including OMB. Both FDA and FSIS have created executive positions to focus agency efforts on food safety, and officials from both agencies told us that the FSWG has increased interagency collaboration. However, while creating the FSWG is a positive first step, we have reported that the continuity of food safety coordination efforts can be hampered by changes in executive branch leadership.[11] As a presidentially appointed working group, the FSWG's future is uncertain, and the experience of the former President's Council on Food Safety, which disbanded less than 3 years after it was created, illustrates that this type of approach can be short lived.

Nevertheless, through the FSWG, federal agencies have taken steps designed to increase collaboration in some areas that cross regulatory jurisdictions—in particular, improving produce safety, reducing *Salmonella* contamination, and developing food safety performance measures.

- *Produce safety.* Preventing contamination of fresh produce is an FSWG priority. In 2009, FDA issued draft guidance for industry on produce safety to minimize food safety hazards and contamination of leafy greens, tomatoes, and melons. FDA has also publicly announced that it is developing a proposed regulation setting enforceable standards for fresh produce safety at farms and packing houses, and FDA and AMS officials told us the agencies are collaborating on the rulemaking process, as AMS establishes quality and condition standards the food industry can voluntarily adopt for marketing purposes through marketing agreements. Specifically, an AMS employee who USDA describes as having extensive experience working with the produce industry for over 20 years has been temporarily assigned to work as a Senior Policy Analyst in FDA's Office of Foods from October 2009 to March 2011 to help develop the regulation. The AMS employee told us she is providing FDA with information about practices on farms and in packing houses that FDA officials are using to inform their rulemaking. She is also helping FDA leverage existing AMSindustry relationships to conduct outreach to farms and packing houses before the produce safety regulation goes into effect. In addition, AMS has publicly announced that it is developing a proposed national marketing agreement for leafy greens that will include food safety standards. The purpose of the proposed marketing agreement is to enhance the quality of fresh leafy green vegetable products through the application of good agricultural and handling practices, and to improve consumer confidence, among other things. Based on our review of *Federal Register* documents and interviews with AMS and FDA officials, we found that AMS is coordinating with FDA to ensure that the standards in the proposed marketing agreement are consistent with the standards FDA is setting in the produce safety regulation.

- *Salmonella contamination.* According to FSIS, FDA, and OMB officials, FSIS and FDA worked together to establish complementary performance goals under the High Priority Performance Goal initiative—a White House management initiative—for reducing illness caused by *Salmonella*. FSIS officials told us that staff from FSIS and FDA communicated on a regular basis to coordinate efforts to develop their respective agencies' goals, as they are closely intertwined. *Salmonella* contamination can occur in poultry and egg products, which are under FSIS's regulatory jurisdiction, and shell eggs, which are primarily under FDA's jurisdiction. Both agencies set goals to reduce illness from *Salmonella* within their own areas of egg safety jurisdiction by the end of 2011. According to the FSIS officials, FSIS and FDA coordinated on ensuring that the goals complemented one another, utilized the same datasets, and covered the same time period so that the agencies measure their progress consistently. FSIS's goal is to reduce the rate of illness due to *Salmonella* in FSISregulated products to 5.3 cases per 100,000 people by 2011, a reduction of approximately 22,600 illnesses below the current baseline with an associated cost reduction of $404 million. FDA's goal is a 10 percent decrease by 2011 over the 2007-2009 average baseline rate of *Salmonella Enteritidis* illness in the population.[12] The agencies have taken additional actions to address *Salmonella*. Specifically, in May 2010, FSIS announced new performance standards that poultry establishments must meet for *Salmonella* and released a compliance guide for industry. In July 2009, FDA issued a final rule requiring shell egg producers to implement measures to prevent *Salmonella Enteritidis* from contaminating eggs and, in August 2010, issued draft guidance for producers on implementing the rule.

- *Food safety performance measures*. FSWG members have proposed 21 performance measures for assessing the federal government's progress toward meeting its crosscutting food safety goals. For example, to assess progress in preventing harm to consumers from unsafe food, one of the performance measures the FSWG is proposing is to track the prevalence of selected foodborne hazards in key commodity groups. As a next step, FSIS and FDA are beginning to set quantitative targets for the measures. For example, one of FSIS's proposed quantitative targets is to reduce the prevalence of *Salmonella Enteritidis* in FSIS prepasteurized egg products from 14.12 percent (the fiscal year 2009 baseline) to 12.71 percent by fiscal year 2015. FDA is developing its own quantitative targets. FSWG and agency documents show that both agencies, with support from CDC, are coordinating to pilot-test a framework for developing the quantitative targets for *Salmonella Enteritidis* that covers multiple agencies' oversight jurisdictions. FSIS officials told us this framework could identify data gaps and help target areas where more attention is needed. An FDA official told us that the agencies plan to use this framework to assess progress toward other food safety goals, but could not estimate when the agencies will finish developing a complete set of quantitative targets for the food safety performance measures.

In addition, we found that federal food safety agencies are taking actions in two other areas. First, FDA is leading an effort involving other federal agencies including CDC, USDA, and the Department of Homeland Security, to increase the inspection, laboratory, and outbreak-response capacities of state and local food safety agencies. FDA issued a draft paper describing its vision and established working groups of federal, state, and local government representatives to address issues such as national standards for food regulatory programs and standardized laboratory practices and procedures, among others. It has also hosted two national meetings that were attended by public health and food safety officials from federal, state, local, and territorial government agencies. Second, in February 2011, federal food safety agencies established a Multi-Agency Coordination group intended to improve the response to outbreaks of foodborne illness, an FSWG priority. According to agency documents, the group will be used to coordinate the response of federal agencies, and state, local, and tribal governments, in managing large-scale foodborne illness outbreaks, prioritizing the allocation of critical resources, and making policy decisions. It is co-chaired by USDA and the Department of Health and Human Services and participating federal agencies include FDA, CDC, FSIS, AMS, the Food and Nutrition Service, the Animal and Plant Health Inspection Service, the Foreign Agricultural Service, the Environmental Protection Agency, and the Departments of Homeland Security, State, and Justice.

However, while the FSWG has taken steps to increase interagency collaboration on food safety, it has not developed a governmentwide performance plan that provides a comprehensive picture of the federal government's food safety efforts. When we added food safety oversight to our high-risk list in 2007, we said that what remains to be done is to develop a governmentwide performance plan for food safety that is mission based, results oriented, and provides a cross-agency perspective.[13] We also said this plan could be used to help decision makers balance trade-offs and compare performance when resource allocation and restructuring decisions are made. Officials from OMB, FDA, and USDA told us the FSWG's July 2009 "key findings" represent the governmentwide plan for food safety.[14] The key findings identify the FSWG's three core principles for improving the safety of the U.S. food supply—prioritizing prevention, strengthening surveillance and enforcement, and improving response and recovery—and outline a number of goals and actions the agencies are taking, or plan to take, to improve food safety. The key findings are mission based and offer a cross-agency perspective. For example, both FSIS and FDA are contributing to goals for reducing illness from *Salmonella and E. coli*, and multiple federal agencies are contributing to goals for improving the response to outbreaks of foodborne illness.

Some of the goals are results oriented, such as FSIS's goal to have 90 percent of poultry establishments meeting its new standards to reduce

Salmonella in turkeys and poultry by the end of 2010, and FDA's goal to reduce foodborne illness from shell eggs by approximately 60 percent by issuing a final rule to control *Salmonella* during egg production. However, most of the goals are not results oriented and do not include performance measures, focusing instead on specific actions the agencies plan to take in the near term. For example, to reduce illness from *E. coli*, FSIS's goal is to issue improved instructions to its workforce and increase its sampling. It is not clear the extent to which these one-time actions will help reduce illness from *E. coli*.

Further, it is not clear how the key findings align with the 21 performance measures for food safety proposed by the FSWG. Our prior work has identified results oriented goals and performance measures as standard elements of performance plans.[15] Because the FSWG's key findings generally lack results oriented goals and performance measures, they do not provide a concrete statement of the federal government's expected performance for food safety that can be used for subsequent comparison with its actual performance. Identifying performance gaps can help decision makers target scarce resources. In addition, the key findings do not include information about the resources that are needed to achieve the FSWG's goals. A discussion of the strategies and resources needed to achieve annual goals is also a standard element of performance plans. The key findings also do not address the entire food supply; for example none of the goals specifically addresses food imports, which represent 60 percent of fresh fruits and vegetables and 80 percent of seafood.

Although performance plans are to be updated on an annual basis, OMB officials told us there are currently no plans to update the key findings, which were issued in 2009. Those officials also told us that, instead, the intent was to integrate the FSWG's planning into the agencies' budgeting processes, which include developing performance goals. While individual agency documents provide important and useful information, they do not provide a broader and more integrated picture of food safety oversight throughout the federal government. Without a governmentwide performance plan for food safety, decision makers do not have a comprehensive picture of the federal government's performance on this crosscutting issue. Further, performance plans provide an opportunity for agencies to identify factors that influence the accomplishment of their goals and discuss the strategies they plan to take to leverage or mitigate the influence such factors can have on achieving results. For example, a governmentwide plan for food safety could recognize that the federal agencies' food safety oversight authorities differ and recognize and address other external factors, such as the actions of state and local governments, which influence the accomplishment of federal food safety goals.

Recent legislation reinforces the need for and importance of governmentwide planning. The GPRA Modernization Act of 2010, which updates the requirements for strategic plans and performance plans, also recognizes the importance of governmentwide planning on crosscutting issues.[16] In addition, the recently enacted FDA Food Safety Modernization Act calls for coordination among federal agencies. Specifically, the Act directs the Secretary of Health and Human Services to improve coordination and cooperation with the Secretaries of Agriculture and Homeland Security to target food inspection resources and to submit annual reports to Congress describing those efforts and providing other information about FDA's inspections. The Act also requires the Secretaries of Health and Human Services and Agriculture, in consultation with the Secretary of Homeland Security, to prepare a National Agriculture and Food Defense Strategy. It also requires the Secretary of Health and Human Services, in coordination with the Secretaries of Agriculture and Homeland Security, to submit a comprehensive report to Congress that identifies programs and practices that are intended to promote the safety and security of the food supply and prevent foodborne outbreaks. The information provided in those reports could help inform a governmentwide performance plan for food safety. Moreover, we have reported that establishing mutually reinforcing or joint strategies is a key practice that can help enhance and sustain interagency collaboration among federal agencies.[17] Such strategies help in aligning the collaborating agencies' activities, core processes, and resources to accomplish a common outcome. In this current fiscal environment, we and others have called for agencies to leverage scarce resources for food safety and other issues.

Our past work on other interagency planning may provide models for governmentwide planning on food safety. While FDA and USDA have collaborated on setting food safety goals for Healthy People—a multiagency initiative to improve public health—they are long-term goals, set every 10 years, and are not linked to resources. We have reported on interagency planning for sharing health resources for military service members and veterans and managing wildland fires.

Health resource sharing. The Department of Veterans Affairs (VA) and Department of Defense (DOD) collaborate to use federal health resources, such as space in medical facilities, under the Veterans Administration and Department of Defense Health Resources Sharing and Emergency Operations Act. In addition, the VA/DOD Joint Executive Council develops a joint strategic plan to shape, focus, and prioritize the coordination and sharing efforts, as directed by Congress and recommended by a presidential task force. The council has developed a new joint strategic plan each year since 2003, which is included in the council's annual report to the VA and DOD Secretaries on the status of implementing its collaboration and sharing activities. We reported that, according to DOD, the joint strategic plan outlines actionable objectives, assigns accountability, and establishes performance targets.[18]

- *Wildland fire management.* Five federal agencies that share responsibility for wildland fire management—the Forest Service at USDA and the Bureau of Indian Affairs, Bureau of Land Management, Fish and Wildlife Service, and National Park Service at the Department of Interior—have long coordinated their fire suppression efforts. The intergovernmental Wildland Fire Leadership Council, which was recommended by GAO and established by the Secretaries of Agriculture and the Interior in 2002, seeks to support implementation of federal fire management policy by coordinating agency policies and providing strategic direction.[19] In reports dating back to 1999, we have recommended that the agencies develop a cohesive strategy that identifies options and funding for preventing and responding to fires.[20] At the direction of Congress, the council began developing a cohesive wildland fire management strategy in 2010 that is required to address fire suppression, prevention, and resource allocation issues.

We and Others Have Identified Options to Reduce Fragmentation and Overlap in Food Safety Oversight, but They Have Not Been Analyzed in Detail

We, the National Academy of Sciences, the Produce Safety Project, and the former President's Council on Food Safety have identified options to reduce fragmentation and overlap in food safety oversight in the form of alternative organizational structures (see app. III), but a detailed analysis of their advantages and disadvantages and the potential challenges that could arise if they are implemented has yet to be conducted.[21] In 2001, we first suggested that Congress consider commissioning the National Academy of Sciences or a blue ribbon panel to conduct a detailed analysis of alternative organizational structures for food safety[22] and reiterated the suggestion over the years, most recently in the 2011 high-risk list update.[23]

Some of the alternative organizational structures that we and others have identified include:

- *Single food safety agency.* All aspects of food safety at the federal level could be consolidated into a single food safety agency, either housed within an existing agency or established as an independent entity. This consolidation would bring oversight of all foods under a single administrator and consolidate tasks that are currently dispersed throughout multiple federal agencies, such as inspections, risk assessment, standard setting, research, and surveillance.
- *Food safety inspection agency.* Food safety inspection activities, but not other activities such as surveillance, could be consolidated under USDA or FDA. As we reported in the past, any new inspection system should employ a unified risk-based approach, which would require Congress to modify the current legislative structure.[24]
- *Data collection and risk analysis center.* Data collection and risk analysis could be consolidated into a single center that would disseminate the results of its analyses to the food safety agencies. For example, this center could consolidate food safety surveillance data collected from a variety of sources and analyze it at the national level to support risk-based decision making. While the center would be independent from the regulatory agencies to give its analyses scientific credibility, it would also consult with the agencies to understand their needs, but would not preempt any agency's authority to develop its own food safety management approach.
- *Coordination mechanism.* Centralized, executive leadership could be provided for the existing organizational structure using a coordination mechanism with representatives from the agencies, similar to the FSWG. However, unlike the FSWG, the coordination mechanism would be led by a central chair who would be appointed by the President and have control over resources.

While a detailed analysis of the alternatives has not been conducted, organizations have offered some preliminary observations on some of their benefits. For example, in its strategic plan, the former President's Council on

Food Safety stated that consolidation could eliminate duplication and fragmentation, create a centralized leadership, clarify lines of authority, and facilitate priority setting and resource allocation based on risk. Similarly, in its 2010 report *Enhancing Food Safety: The Role of the Food and Drug Administration*, the National Academy of Sciences concluded that the core federal food safety responsibilities should reside within a single entity having a unified administrative structure, clear mandate, dedicated budget, and full responsibility for oversight of the entire food supply. In its report, the National Academy of Sciences also stated that centralizing data collection and risk analysis would eliminate the need for each agency to develop its own comprehensive expertise in risk and decision analysis; promote communication, collaboration, and data sharing among federal agencies; and could be a first step toward accomplishing the more challenging goal of consolidating all federal food safety activities into a single agency.

We recognize that reorganizing federal food safety responsibilities would be a complex process. Further, our work on other agency mergers and transformations indicates that reorganizing food safety could have short-term disruptions and transition costs.[25] We reported that a merger or transformation is a substantial commitment that could take years before it is completed, and therefore must be carefully and closely managed. In particular, the experience of major private sector mergers and acquisitions has been that productivity and effectiveness actually decline initially, in part because attention is concentrated on critical and immediate integration issues and diverted from longer-term mission issues. Our work on seven other countries' experiences between 1997 and 2004 in consolidating their food safety systems found that while the extent to which those countries consolidated their food systems varied considerably, they faced similar challenges in deciding whether to place the new agency within the existing health or agriculture agency or establish it as a stand-alone agency and in determining what responsibilities the new agency would have.[26] We also reported that the countries experienced benefits, such as improved public confidence in their food safety systems.[27] In addition, each country modified its existing legal framework to give legal authority and responsibility to the new food safety agency. Some European Union (EU) countries were further prompted to consolidate in order to comply with new EU food safety legislation that became effective, in large part, in January 2006. The EU adopted comprehensive food safety legislation in 2004 intended to create a single, transparent set of food safety rules applicable to both animal and nonanimal products.

We and other organizations have regularly paired proposals for alternative food safety organizations with calls for comprehensive, unified, risk-based food safety legislation.

Existing statutes give agencies different regulatory and enforcement authorities, and we have reported that legislation governing the agencies' authorities, jurisdictions, and inspection frequencies is not the product of strategic design as to how to protect public health.[28]

In May 2004, we reported that a critical step in designing and implementing a risk-based food safety system is identifying the most important food safety problems across the entire food system from a public health perspective and concluded that comprehensive, uniform, and risk-based food safety legislation is needed to provide the foundation for this approach.[29]

The National Academy of Sciences also concluded that to create a science-based food safety system current laws must be revised, and recommended that Congress change federal statutes so that inspection, enforcement, and research efforts can be based on risks to public health. While the new food safety law strengthens a major part of the food safety system and expands FDA's oversight authority, it does not apply to the federal food safety system as a whole or create a new risk-based food safety structure.

In February 2011, we reiterated our suggestion for comprehensive, unified, risk-based food safety legislation.[30]

CONCLUSION

We are encouraged by the executive branch's attention to food safety through the FSWG and its initial efforts designed to improve interagency collaboration on this very important crosscutting issue. However, food safety remains fragmented and much work remains to be done on several of the FSWG initiatives.

Further, the collaboration that has begun under the FSWG may be short lived, putting some of the longer-term efforts, such as developing results-oriented food safety goals and measures, at risk of not being completed. Because food safety oversight faces ongoing challenges, it is important that this issue be given sustained attention; as we have previously reported, the continuity of food safety coordination efforts can be hampered when executive branch leadership changes.

Thus, it is critical that the primary food safety agencies engage in comprehensive, governmentwide planning to increase interagency collaboration under the current system. Such a plan, paired with comprehensive risk-based food safety legislation and a detailed analysis of alternative organizational structures for food safety oversight, could be an important tool for addressing fragmentation in federal food safety oversight.

However, without an annually updated governmentwide performance plan for food safety that contains results-oriented goals and performance measures and a discussion of strategies and resources used by the agencies with food safety responsibilities, decision makers do not have a comprehensive picture of the federal government's performance on this crosscutting issue.

Further, without such information, decision makers may be hampered in their efforts to make key resource allocation and restructuring decisions.

Recommendation for Executive Action

In order to improve collaboration among federal agencies on food safety oversight and provide an integrated perspective on this crosscutting issue we are making the following recommendation.

The Director of the Office of Management and Budget, in consultation with the federal agencies that have food safety responsibilities, should develop a governmentwide performance plan for food safety.

The performance plan should include results-oriented goals and performance measures for food safety oversight throughout the federal government, as well as a discussion about strategies and resources. It should be updated on an annual basis.

Agency Comments

We provided USDA, the Department of Health and Human Services, and OMB with drafts of this report for review. OMB declined to comment on the draft report. USDA and Health and Human Services provided technical comments, which we incorporated as appropriate.

Lisa Shames
Director, Natural Resources and Environment

APPENDIX I: SCOPE AND METHODOLOGY

A new statutory requirement mandates that GAO identify programs, agencies, offices, and initiatives with duplicative goals and activities within departments and governmentwide.[1]

Under that mandate this review examines:

(1) steps, if any, that the Food Safety Working Group (FSWG) has taken to increase collaboration among federal food safety agencies, and

(2) options we and others have identified to reduce fragmentation, overlap, and potential duplication in food safety oversight.

To complete our work we reviewed food safety reports and legislation, and interviewed officials from the Department of Agriculture (USDA), the Food and Drug Administration (FDA), and the Office of Management and Budget (OMB).

To address the first question we also collected and analyzed information about the FSWG, its activities, and its plan for food safety. We also collected documentation about the FSWG's activities from the agencies, the Federal Register, and budget documents.

We assessed the FSWG's "key findings," which FSWG officials told us represent the governmentwide plan for food safety, against GAO's criteria for performance plans.[2]

To identify options for reducing fragmentation, overlap, and potential duplication, we identified alternative organizational structures for food safety by reviewing reports by GAO, the National Academy of Sciences, the Produce Safety Project, and the former President's Council on Food Safety.

We also reviewed reports by GAO about federal agency and private sector mergers and organizational transformations. We did not independently verify the foreign laws discussed in this report.

We conducted this performance audit from July 2010 to March 2011 in accordance with generally accepted government auditing standards.

Those standards require that we plan and perform the audit to obtain sufficient, appropriate evidence to provide a reasonable basis for our findings and conclusions based on our audit objectives.

We believe that the evidence obtained provides a reasonable basis for our findings and conclusions based on our audit objectives.

APPENDIX II: FEDERAL AGENCIES WITH FOOD SAFETY RESPONSIBILITIES

	Agency	Programs	Regulation	Inspection	Enforcement	Research	Grant & Cooperative Agreements Admin.	Other	Responsibilities and main authorizing statutes
1	USDA Food Safety and Inspection Service	Meat, Poultry, Egg Products, and Catfish Inspection; Voluntary Fee for Service Inspections	●	●				●	• **Responsible for:** Ensuring that the nation's domestic and imported commercial supply of meat, poultry, egg products, and catfish[a] is safe, wholesome, and correctly labeled and packaged, and for enforcing the Humane Methods of Slaughter Act of 1978, as amended. Responsible for providing voluntary fee-for-service inspections for exotic and other edible animals. • **Main authorizing statutes:** Poultry Products Inspection Act, Pub. L. No. 85-172, 71 Stat. 441 (1957) (codified as amended at 21 U.S.C. ss. 451-472); Federal Meat Inspection Act, ch. 2907, 34 Stat. 1256, 1260 (1907) (codified as amended at 21 U.S.C. ss. 601-695); Egg Products Inspection Act, Pub. L. No. 91-597, 84 Stat. 1620 (1970) (codified as amended at 21 U.S.C. ss. 1031-1056). See also, Humane Methods of Slaughter Act of 1978, Pub. L. No. 95-445, 92 Stat. 1069 (codified as amended at 7 U.S.C. ss. 1902, 1904, 21 U.S.C. ss. 603, 610, 620); Federal Anti-Tampering Act, Pub. L. No. 98-127, s. 2, 97 Stat. 831, 831 (1983) (codified as amended at 18 U.S.C. s. 1365); Agricultural Marketing Act of 1946, ch. 966, tit. II, s. 203, 60 Stat. 1087, 1087 (codified as amended at 7 U.S.C. s. 1622). See also National School Lunch Act, ch. 281, 60 Stat. 230 (1946) (codified as amended at 42 U.S.C. ss. 1751-1770), as amended by Child Nutrition and WIC Reauthorization Act of 2004, Pub. L. No. 108-265, s. 118, 118 Stat. 729, 752 (codified as amended at 42 U.S.C. s. 1762a(h)).
2	USDA Animal and Plant Health Inspection Service	Plant Protection and Quarantine; Veterinary Services	●	●					• **Responsible for:** Preventing the introduction or dissemination of plant pests. Responsible for preventing the introduction or dissemination of livestock pests or diseases. • **Main authorizing statutes:** Agricultural Bioterrorism Act of 2002, Pub. L. No. 107-188, tit. II, subtit. B, 116 Stat. 647 (codified as amended at 7 U.S.C. s. 8401); Animal Health Protection Act, Pub. L. No. 107-171, tit. X, 116 Stat. 494 (2002) (codified as amended at 7 U.S.C. ss. 8301-8322); Plant Protection Act, Pub. L. No. 106-224, tit. IV, 114 Stat. 438 (2000) (codified as amended at 7 U.S.C. ss. 7701-7721).
3	USDA Grain Inspection, Packers and Stockyards Administration	Federal Grain Inspection Service	●	●	●				• **Responsible for:** Establishing quality standards, inspection procedures, and marketing of grain and other related products. • **Main authorizing statutes:** United States Grain Standards Act, ch. 313, 39 Stat. 482 (1916) (codified as amended at 7 U.S.C. ss. 71-87k).

Appendix II. (Continued)

	Agency	Programs	Regulation	Inspection	Enforcement	Research	Grant & Cooperative Agreements Admin.	Other	Responsibilities and main authorizing statutes
4	USDA Agricultural Marketing Service	Commodity Programs; Science and Technology Programs; School Lunch Commodity Purchases	●	●	●			●	• **Responsible for:** Establishing quality and condition standards for, among other things, dairy, fruit and vegetables, livestock. • **Main authorizing statutes:** Agricultural Marketing Act of 1946, ch. 966, tit II, 60 Stat. 1087 (codified as amended at 7 U.S.C. ss. 1621-1638d). See also e.g., Perishable Agricultural Commodities Act, 1930, ch. 436, 46 Stat. 531 (codified as amended at 7 U.S.C. s. 499a- 499s); Federal Seed Act, Ch. 615, 53 Stat. 1275 (1939) (codified as amended at 7 U.S.C. ss. 1551-1611).
5	USDA Agricultural Research Service	Nutrition, Food Safety, and Quality				●	●		• **Responsible for:** Providing the scientific research to help ensure that the food supply is safe and secure and that foods meet foreign and domestic regulatory requirements. • **Main authorizing statutes:** 7 U.S.C. ss. 1622, 2204, 3101, 3121, 3318, 3319a; also e.g., 7 U.S.C. ss. 136i-2, 391, 7654.
6	USDA Economic Research Service	Food Safety Research				●	●		• **Responsible for:** Providing analyses of the economic issues affecting the safety of the U.S. food supply. • **Main authorizing statutes:** 7 U.S.C. ss. 1622, 2204, 3101, 3121, 3318, 3319a; also e.g., 7 U.S.C. ss. 136i-2, 391, 7654.
7	USDA National Agricultural Statistics Service	Statistical Program on Food Safety				●	●		• **Responsible for:** Providing statistical data, including agricultural chemical usage data, related to the safety of the food supply. • **Main authorizing statutes:** 7 U.S.C. ss. 1622, 2204, 3101, 3121, 3318, 3319a; also e.g., 7 U.S.C. ss. 136i-2, 391, 7654.
8	USDA National Institute of Food and Agriculture	National Integrated Food Safety Research Initiative	●				●		• **Responsible for:** Supporting food safety projects in the land-grant university system and other partner organizations that demonstrate an integrated approach to solving problems in applied food safety research, education, or extension. • **Main authorizing statutes:** 7 U.S.C. ss. 361a-361i, 3121, 3151, 3155, 3318, 3319a, 6971(f); also e.g., 7 U.S.C. ss. 450i, 3902.
9	HHS Food and Drug Administration	Foods Program; Animal Drugs and Feeds Program; Regional Operations and Enforcement	●	●	●	●	●	●	• **Responsible for:** Ensuring that all domestic and imported foods, excluding meat and poultry products, are safe, wholesome, sanitary, and properly labeled. • **Main authorizing statutes:** Federal Food, Drug, and Cosmetic Act, ch. 675, 52 Stat. 1040 (1938) (codified as amended at 21 U.S.C. ss. 301-399a), as amended by, among others, Food Additives Amendment of 1958, Pub. L. No. 85-929, 72 Stat. 1784; Safe Drinking Water Act of 1974, Pub. L. No. 93-523, 88 Stat. 1660; Infant Formula Act of 1980, Pub. L. No. 96-359, 94 Stat. 1190; Nutrition Labeling and Education Act of 1990, Pub. L. No. 101-535, 104 Stat. 2353; Dietary Supplement Health and Education

Agency	Programs	Regulation	Inspection	Enforcement	Research	Grant & Cooperative Agreements Admin.	Other	Responsibilities and main authorizing statutes
9 Food and Drug Administration continued								Act of 1994, Pub. L. No. 103-417, 108 Stat. 4325; Food and Drug Modernization Act of 1997, Pub. L. No. 105-115, 111 Stat. 2296; Public Health Security and Bioterrorism Preparedness and Response Act of 2002, Pub. L. No. 107-188, tit. III, 116 Stat. 594; Sanitary Food Transportation Act of 2005, Pub. L. No. 109-59, tit. VII, s. 7202, 119 Stat. 1891, 1911; FDA Food Safety Modernization Act, Pub. L. No. 111-353, 124 Stat. 3885 (2011). See also, Act of February 15, 1927 (Federal Import Milk Act), ch. 155, 44 Stat. 1101 (codified as Amended at 21 U.S.C. 141-149); Fair Packaging and Labeling Act, Pub. L. No. 89-755, 80 Stat. 1296 (1966) (codified as amended at 15 U.S.C. ss. 1451-1461); Federal Anti-Tampering Act, Pub. L. No. 98-127, s. 2, 97 Stat. 831, 831 (1983) (codified as amended at 18 U.S.C. s. 1365); Pesticide Monitoring Improvements Act of 1988, Pub. L. No. 100-418, s. 4702, 102 Stat. 1411, 1412 (codified as amended at 21 U.S.C. s. 1401).
10 HHS Centers for Disease Control and Prevention	Food Safety Initiative	●	●		●			• **Responsible for:** Preventing the transmission, dissemination, and spread of foodborne illness to protect the public health. • **Main authorizing statutes:** Public Health Service Act, ch. 373, 58 Stat. 682 (1944) (codified as amended at 42 U.S.C. ss. 201-300bbb).
11 Commerce National Marine Fisheries Service	Seafood Inspection Program						●	• **Responsible for:** Providing voluntary, fee-for-service examinations of seafood for safety and quality • **Main authorizing statutes:** Agricultural Marketing Act of 1946, ch. 966, tit. II, s. 203, 205, 60 Stat. 1087, 1087 (codified as amended at 7 U.S.C. ss. 1622, 1624). See also Act of May 25, 1900 (Lacey Act), ch. 553, 31 Stat. 187 (codified as amended in part at 16 U.S.C. s. 3371).
12 EPA	Pollution Prevention and Toxins; Pesticides Program; Safe Drinking Water	●						• **Responsible for:** Regulating the use of certain chemicals and substances that present an unreasonable risk of injury to health or the environment. Responsible for issuing regulations to establish, modify, or revoke tolerances for pesticide chemical residues. Responsible for setting national drinking water standard of quality and consulting with FDA before FDA promulgates regulations for standard of quality for bottled water. • **Main authorizing statutes:** Toxic Substance Control Act, Pub. L. No. 94-469, 90 Stat. 2003 (1976) (codified as amended at 15 U.S.C. ss. 2601-2697) Federal Insecticide, Fungicide, and Rodenticide Act, ch. 125, 61 Stat. 163 (1947) (codified as amended at 7 U.S.C. ss. 136-136y), as amended by the Food Quality Protection Act of 1996, Pub. L. No. 104-170, 110 Stat. 1489; 21 U.S.C. s. 346a, Safe Drinking Water Act of 1974, Pub. L. No. 93-523, 88 Stat. 1660 (codified as amended at 21 U.S.C. s. 349 and 42 U.S.C. ss. 300f through 300j-26).

Appendix II. (Continued)

	Agency	Programs	Regulation	Inspection	Enforcement	Research	Grant & Cooperative Agreements Admin.	Other	Responsibilities and main authorizing statutes
13	Treasury Alcohol, Tobacco, Firearms, and Explosives	Alcohol	●		●			●	• **Responsible for:** Regulating, enforcing, and issuing permits for the production, labeling, and distribution of alcoholic beverages. • **Main authorizing statutes:** Federal Alcohol Administration Act, ch. 814, 49 Stat. 977 (1935) (codified as amended at 27 U.S.C. ss. 201-219a).
14	Department of Homeland Security Customs and Border Protection			●	●				• **Responsible for:** Inspecting imports, including food products, plants, and live animals, for compliance with U.S. law and assisting all federal agencies in enforcing their regulations at the border. • **Main authorizing statutes:** Tariff Act of 1930, 19 U.S.C. ss. 1202-1654. See also Homeland Security Act of 2002, Pub. L. No. 107-296, s. 421, 116 Stat. 2135, 2182.
15	Federal Trade Commission				●				• **Responsible for:** Enforcing prohibitions against false advertising for, among other things, food products. • **Main authorizing statutes:** Federal Trade Commission Act, ch. 311, 38 Stat. 717 (1914) (codified as amended at 15 U.S.C. ss. 41-58).

Source: GAO analysis.

[a] The 2008 Farm Bill amended the Federal Meat Inspection Act to give USDA responsibility for the inspection of catfish. The amendments specified that they would not apply until USDA issues final regulations implementing them, a process that was not yet complete as of February 2011.

End Notes

[1] GAO, *High Risk Series: An Update*, GAO-07-310 (Washington, D.C.: January 2007).

[2] GAO, *21st Century Challenges: Reexamining the Base of the Federal Government*, GAO-05-325SP (Washington, D.C.: February 2005).

[3] GAO, *Oversight of Food Safety Activities: Federal Agencies Should Pursue Opportunities to Reduce Overlap and Better Leverage Resources*, GAO-05-213 (Washington, D.C.: Mar. 30, 2005).

[4] GAO, *Food Safety and Security: Fundamental Changes Needed to Ensure Safe Food*, GAO-02-47T (Washington, D.C.: Oct. 10, 2001).

[5] GAO-07-310.

[6] GAO-02-47T.

[7] Pub. L. No. 111-353, 124 Stat. 3885 (2011).

[8] Pub. L. No. 111-139 § 21, 124 Stat. 8, 29 (2010).

[9] *Budget of the U.S. Government, Fiscal Year 2011* (Washington, D.C.: February 2010). Fiscal year 2009 budget obligations for FDA's Foods Program include funds that were appropriated to FDA in fiscal year 2008 but available for obligation through fiscal year 2009, according to FDA officials. Budget obligations for FDA do not include food safety activities in the Office of the Commissioner through the Office of Foods, or the Animal Drugs and Feed program. Fiscal year 2009 budget obligations for FSIS do not include additional fees FSIS collects from establishments, importers, and exporters to offset costs for overtime inspection services.

[10] In September 2010, USDA officials told us the agency estimated it would have to spend approximately $30 million in fiscal years 2011 and 2012 to develop and implement its catfish inspection program. Because of the need for considerable stakeholder engagement and regulatory development before the program's adoption and implementation, USDA officials revised their estimate for fiscal year 2011 and no funding was proposed for the program in the President's fiscal year 2012 budget.

[11] GAO-02-47T.

[12] According to CDC, *Salmonella Enteritidis* is one of the most common variations of *Salmonella* bacteria reported worldwide.

[13] GAO-07-310.

[14] See key findings at http://www.foodsafetyworkinggroup.gov/.

[15] GAO, *Managing For Results: Using the Results Act to Address Mission Fragmentation and Program Overlap*, GAO/AIMD-97-146 (Washington, DC: Aug. 29, 1997); *Agency Performance Plans: Examples of Practices That Can Improve Usefulness to Decisionmakers*, GAO/GGD/AIMD-99-69 (Washington, D.C.: Feb. 26, 1999); and *Agencies' Annual Performance Plans Under The Results Act: An Assessment Guide to Facilitate Congressional Decisionmaking*, GAO/GGD/AIMD-10.1.18 (Washington, D.C.: February 1998).

[16] GPRA Modernization Act of 2010, Pub. L. No. 111-352, 124 Stat. 3866.

[17] GAO, *Results-Oriented Government: Practices That Can Help Enhance and Sustain Collaboration among Federal Agencies*, GAO-06-15 (Washington, D.C.: Oct. 21, 2005).

[18] GAO, *VA and DOD Health Care: Progress Made on Implementation of 2003 President's Task Force Recommendations on Collaboration and Coordination, but More Remains to be Done*, GAO-08-495R (Washington, D.C.: Apr. 30, 2008).

[19] GAO, *Severe Wildland Fires: Leadership and Accountability Needed to Reduce Risks to Communities and Resources*, GAO-02-259 (Washington, D.C.: Jan. 31, 2002).

[20] GAO, *Western National Forests: A Cohesive Strategy is Needed to Address Catastrophic Wildfire Threats,* GAO/RCED-99-65 (Washington, D.C.: Apr. 2, 1999); *Wildland Fire Management: Important Progress Has Been Made, but Challenges Remain to Completing a Cohesive Strategy*, GAO-05-147 (Washington, D.C.: Jan. 14, 2005, GAO); and *Wildland Fire Management: Federal Agencies Have Taken Important Steps Forward but Additional Strategic Action Is Needed to Capitalize on Those Steps*, GAO-09-877 (Washington, D.C.: Sept. 9, 2009).

[21] The Produce Safety Project is an initiative of the Pew Charitable Trusts at Georgetown University. The President's Council on Food Safety was created by the President in 1998 and disbanded in 2001.

[22] GAO-02-47T.

[23] GAO, *High Risk Series: An Update*, GAO-11-278 (Washington, D.C.: February 2011).

[24] GAO, *Posthearing Questions Related to Fragmentation and Overlap in the Federal Food Safety System*, GAO-04-832R (Washington, D.C: May 26, 2004).

[25] GAO, *Results Oriented Cultures: Implementation Steps to Assist Mergers and Organizational Transformations*, GAO-03-669 (Washington, D.C.: July 2, 2003).

[26] GAO, *Food Safety: Experiences of Seven Countries in Consolidating Their Food Safety Systems*, GAO-05-212 (Washington, D.C.: Feb. 22, 2005).

[27] GAO, *Food Safety: Selected Countries' Systems Can Offer Insights into Ensuring Import Safety and Responding to Foodborne Illness,* GAO-08-794 (Washington, D.C.: June, 10, 2008).

[28] GAO, *Federal Food Safety and Security System: Fundamental Restructuring Is Needed to Address Fragmentation and Overlap*, GAO-04-588T (Washington, D.C.: Mar. 30, 2004).

[29] GAO-04-832R. [30]GAO-11-278.

End Notes for Appendix II

[1] Pub. L. No. 111-139 § 21, 124 Stat. 8, 29 (2010).

[2] GAO, *Managing For Results: Using the Results Act to Address Mission Fragmentation and Program Overlap*, GAO/AIMD-97-146 (Washington, DC: Aug. 29, 1997); *Agency Performance Plans: Examples of Practices That Can Improve Usefulness to Decisionmakers*, GAO/GGD/AIMD-99-69 (Washington, D.C.: Feb. 26, 1999); and *Agencies' Annual Performance Plans Under The Results Act: An Assessment Guide to Facilitate Congressional Decisionmaking*, GAO/GGD/AIMD-10.1.18 (Washington, D.C.: February 1998).

In: The Federal Food Safety Working Group ISBN: 978-1-62417-059-1
Editors: R. Jaynes and S. Darlington

Chapter 3

THE FEDERAL FOOD SAFETY SYSTEM: A PRIMER*

Renée Johnson

SUMMARY

Numerous federal, state, and local agencies share responsibilities for regulating the safety of the U.S. food supply. Federal responsibility for food safety rests primarily with the Food and Drug Administration (FDA) and the U.S. Department of Agriculture (USDA). FDA, an agency of the Department of Health and Human Services, is responsible for ensuring the safety of all domestic and imported food products (except for most meats and poultry). FDA also has oversight of all seafood, fish, and shellfish products. USDA's Food Safety and Inspection Service (FSIS) regulates most meat and poultry and some egg products. State and local food safety authorities collaborate with federal agencies for inspection and other food safety functions, and they regulate retail food establishments.

The combined efforts of the food industry and government regulatory agencies often are credited with making the U.S. food supply among the safest in the world. However, critics view this system as lacking the organization, regulatory tools, and resources to adequately combat foodborne illness—as evidenced by a series of widely publicized food safety problems, including concerns about adulterated food and food

* This is an edited, reformatted and augmented version of Congressional Research Service, Publication No. RS22600, dated January 20, 2012.

ingredient imports, and illnesses linked to various types of fresh produce, to peanut products, and to some meat and poultry products. Some critics also note that the organizational complexity of the U.S. food safety system as well as trends in U.S. food markets—for example, increasing imports as a share of U.S. food consumptions and increasing consumption of fresh, often unprocessed, foods—pose ongoing challenges to ensuring food safety.

The 111th Congress passed comprehensive food safety legislation with the FDA Food Safety Modernization Act (FSMA, P.L. 111-353). FSMA is the largest expansion of FDA's food safety authorities since the 1930s. Although numerous agencies share responsibility for regulating food safety, FSMA focused on foods regulated by FDA and amended FDA's existing structure and authorities, and did not directly address meat and poultry products under USDA's jurisdiction. Beyond these changes, some in Congress continue to push for additional policy reforms to address other perceived concerns about the safety of the U.S. food supply.

BACKGROUND

Americans spend more than $1 trillion on food each year, nearly half of it in restaurants, schools, and other places outside the home.[1] Federal laws give food manufacturers, distributors, and retailers the basic responsibility for assuring that foods are wholesome, safe, and handled under sanitary conditions. A number of federal agencies, cooperating with state, local, and international entities, play a major role in regulating food quality and safety under these laws.

The combined efforts of the food industry and the regulatory agencies often are credited with making the U.S. food supply among the safest in the world. Nonetheless, the Centers for Disease Control and Prevention (CDC) reports that each year an estimated one in six Americans—a total of 48 million people—becomes sick from contaminated food foodborne illnesses caused by contamination from any one of a number of microbial pathogens.[2] Of these, an estimated 128,000 cases require hospitalization and 3,000 cases result in death. In addition, experts have cited numerous other hazards to health, including the use of unapproved veterinary drugs, pesticides, and other dangerous substances in food commodities, of particular concern at a time when a growing share of the U.S. food supply is from overseas sources. These concerns, combined with the ongoing recurrence of major food safety-related incidents, have heightened public and media scrutiny of the U.S. food safety system and magnified congressional interest in the issue.

THE AGENCIES AND THEIR ROLES

The Government Accountability Office (GAO) has identified as many as 15 federal agencies collectively administering at least 30 laws related to food safety. The Food and Drug Administration (FDA), which is part of the U.S. Department of Health and Human Services (HHS), and the Food Safety and Inspection Service (FSIS), which is part of the U.S. Department of Agriculture (USDA), together comprise the majority of both the total funding and the total staffing of the government's food regulatory system. (See Table 1 and Table 2 for a brief comparative look at the agencies' responsibilities.)

Food and Drug Administration

Excluding the roughly 10%-20% of the U.S. food supply that is estimated to be under the jurisdiction of FSIS, it may be argued that the safety of the majority of all foods—about 80%- 90%—falls under FDA's responsibility.[3] The FDA is responsible for ensuring that all domestic and imported food products—except for most meats and poultry—are safe, nutritious, wholesome, and accurately labeled. Examples of FDA-regulated foods are produce, dairy products, seafood, and processed foods. FDA has jurisdiction over meats from animals or birds that are not under the regulatory jurisdiction of FSIS. FDA shares responsibility for the safety of eggs with FSIS. FDA has jurisdiction over establishments that sell or serve eggs or use them as an ingredient in their products. FDA is also responsible for ensuring that most seafood products do not endanger public health (FSIS is to begin inspecting farmed catfish products under a 2008 farm bill provision).

The primary statutes governing FDA's activities are the Federal Food, Drug, and Cosmetic Act (FFDCA), as amended (21 U.S.C. 301 *et seq.*); the Public Health Service Act, as amended (42 U.S.C. §§201 *et seq.*); and the Egg Products Inspection Act, as amended (21 U.S.C. §§1031 *et seq.*). The 111th Congress passed comprehensive food safety legislation with the FDA Food Safety Modernization Act (FSMA, P.L. 111-353). FSMA is the largest expansion of FDA's food safety authorities since the 1930s. Although numerous agencies share responsibility for regulating food safety, FSMA focused on foods regulated by FDA and amended FDA's existing structure and authorities, in particular FFDCA.[4] The Congressional Budget Office (CBO) estimated that implementing FSMA could increase net federal spending subject to appropriation by about $1.4 billion over a five-year period (FY2011-FY2015); collections from possible revenue and direct spending increases from new criminal penalties would be "insignificant, yielding a negligible net impact in each year."[5] FSMA authorizes an increase in FDA staff to up to 5,000 staff members in FY2014. Given the current budgetary climate, funding to undertake many federal activities in FSMA is uncertain.[6]

FDA's FY2011 budget for its foods program was $835.7 million, virtually all of it appropriated with limited authorized user fees.[7] FSMA also authorized additional user fees that provide limited funds to assist with implementation of the new law. FDA staff working on food-related activities in FY2011 were estimated at about 3,400 FTEs.

The number of food facilities subject to FDA inspection has been increasing sharply in the past few years. A report by HHS's Office of Inspector General (OIG) provided additional insights into the FDA's inspections of domestic facilities. The OIG reported that the number of facilities subject to such inspections had risen from about 59,000 in 2004 to nearly 68,000 in 2008. However, the number of inspections conducted declined from about 17,000 in 2004 (29% of the total) to about 15,000 in 2008 (22%). During the five-year period examined by the OIG, 56% of food facilities were not inspected at all.[8] Of these inspections, an increasing number of inspections have been conducted by states under contract with FDA, rather than by FDA.[9]

In addition, some 200,000 foreign food facilities are registered with the agency. Various estimates of unannounced compliance inspections of domestic establishments by FDA officials range from once every five years to once every 10 years, on average, although the agency claims to visit about 6,000 so-called high-risk facilities on an annual basis. FDA relies on notifications from within the industry or from other federal or state inspection personnel, as well as other sources, to alert it to situations calling for increased inspection. GAO reported that, in 2000, FDA inspections covered only about 1% of the food imported under its jurisdiction.[10] Changes to FDA's import regime now being implemented under FSMA are expected to address some of these concerns.

In the Washington, DC, area, two FDA offices are the focal point for food safety-related activities. The Center for Food Safety and Applied Nutrition (CFSAN) is responsible for (1) conducting and supporting food safety research; (2) developing and overseeing enforcement of food safety and quality regulations; (3) coordinating and evaluating FDA's food surveillance and compliance programs; (4) coordinating and evaluating cooperating states' food safety activities; and (5) developing and disseminating food safety and regulatory information to consumers and industry. FDA's Center for Veterinary Medicine (CVM) is responsible for ensuring that all animal drugs, feeds (including pet foods), and veterinary devices are safe for animals, are properly labeled, and produce no human health hazards when used in food-producing animals.

The FDA also cooperates with over 400 state agencies across the nation that carry out a wide range of food safety regulatory activities. However, the state agencies are primarily responsible for actual inspection. FDA works with the states to set the safety standards for food establishments and commodities and evaluates the states' performance in upholding such standards as well as any federal standards that may apply. FDA also contracts with states to use their food safety agency personnel to carry out certain field inspections in support of FDA's own statutory responsibilities.

Food Safety and Inspection Service

Available USDA data indicate that FSIS is roughly responsible for 10%-20% of the U.S. food supply, while FDA is responsible for the remainder.[11] FSIS regulates the safety, wholesomeness, and proper labeling of most domestic and imported meat and poultry and their products sold for human consumption.

Under the Federal Meat Inspection Act of 1906, as amended (21 U.S.C. §§601 *et seq.*), FSIS is required to inspect all cattle, sheep, swine, goats, and equines during slaughtering and processing. Under the Poultry Products Inspection Act of 1957, as amended (21 U.S.C. §§451 *et seq.*), FSIS is required to inspect "any domesticated bird" being processed for human consumption; however, USDA regulations implementing this law limit the definition of domesticated birds to chickens, turkeys, ducks, geese, ratites (emus, ostriches, and rheas), and guineas. FDA has jurisdiction over exotic and alternative meats not inspected by FSIS, and shares the responsibility for egg safety with FSIS.[12] The latter is responsible for the safety of liquid, frozen, and dried egg products, domestic and imported, and for the safe use or disposition of damaged and dirty eggs under the Egg Products Inspection Act, as amended (21 U.S.C. §§1031 *et seq.*).

FSIS's FY2011 budget was $1.007 billion in appropriated funds, plus another approximately $150 million in industry-paid user fees.[13] FSIS staff numbers around 9,600; roughly 8,000 of them, including about 1,000 veterinarians, are in about 6,300 meat slaughtering and/or processing plants nationwide. FSIS personnel inspect all meat and poultry animals at slaughter on a continuous basis, and at least one federal inspector is on the line during all hours the plant is operating. Processing inspection does not require an FSIS inspector to remain constantly on the production line or to inspect every item. Instead, inspectors are on site daily to monitor the plant's adherence to the standards for sanitary conditions, ingredient levels, and packaging, and to conduct statistical sampling and testing of products. Because all plants are visited daily, processing inspection also is considered to be continuous.

FSIS also is responsible for certifying that foreign meat and poultry plants are operating under an inspection system equivalent to the U.S. system before they can export their product to the United States. FSIS inspectors located at U.S. ports of entry carry out a statistical sampling program to verify the safety of imported meats from cattle, sheep, swine, goats, and equines and imported poultry meat from chickens, turkeys, ducks, geese, quail, ratites, and guineas before they are released into domestic commerce. FDA is responsible for ensuring the safety of imported meat from any other species.

Twenty-seven states operate their own meat and/or poultry inspection programs. FSIS is statutorily responsible for ensuring that the states' programs are at least equal to the federal program.[14] Plants processing meat and poultry under state inspection can market their products only within the state. If a state chooses to discontinue its own inspection program, or if FSIS determines that it does not meet the agency's equivalency standards, FSIS must assume the responsibility for inspection if the formerly state-inspected plants are to remain in operation. FSIS also has cooperative agreements with more than two dozen states under which state inspection personnel are authorized to carry out federal inspection in meat and/or poultry plants. Products from these plants may travel in interstate commerce.[15]

Centers for Disease Control and Prevention

CDC is responsible for (1) monitoring, identifying, and investigating foodborne disease problems to determine the contributing factors; (2) working with FDA, FSIS, NMFS, state and local public health departments, universities, and industry to develop control methods; and (3) evaluating the effect of control methods. In 1995, CDC launched "FoodNet," a collaborative project with the FDA and USDA to improve data collection on foodborne illness outbreaks. FoodNet includes active surveillance of clinical microbiology laboratories to obtain a more accurate accounting of positive test results for foodborne illness; a physician survey to determine testing and laboratory practices; population surveys to identify illnesses not reported to doctors; and research studies to obtain new and more precise information about which food items or other exposures may cause diseases. FoodNet data allows CDC to have a clearer picture of the incidence and causes of foodborne illness and to establish baseline data against which to measure the success of changes in food safety programs. The Public Health Service Act provides legislative authority for CDC's food safety-related activities.

National Marine Fisheries Service

Although the FDA is the primary agency responsible for ensuring the safety, wholesomeness, and proper labeling of domestic and imported seafood products, the National Marine Fisheries Service (NMFS), which is part of the U.S. Department of Commerce, conducts, on a fee-forservice basis, a voluntary seafood inspection and grading program that focuses on marketing and quality attributes of U.S. fish and shellfish. The primary legislative authority for NMFS's inspection program is the Agricultural Marketing Act of 1946, as amended (7 U.S.C. §§1621 *et seq.*). NMFS has approximately 160 seafood safety and quality inspectors, and inspection services are funded with user fees.

Environmental Protection Agency

EPA has the statutory responsibility for ensuring that the chemicals used on food crops do not endanger public health. EPA's Office of Pesticide Programs is the part of the agency that (1) registers new pesticides and determines residue levels for regulatory purposes; (2) performs special reviews of pesticides of concern; (3) reviews and evaluates all the health data on pesticides; (4) reviews data on pesticides' effects on the environment and on other species; (5) analyzes the costs and benefits of pesticide use; and (6) interacts with EPA regional offices, state regulatory counterparts, other federal agencies involved in food safety, the public, and others to keep them informed of EPA regulatory actions. The Federal Insecticide, Fungicide, and Rodenticide Act, as amended (7 U.S.C. 136 et seq.), and the Federal Food, Drug, and Cosmetic Act, as amended (21 U.S.C. 301 et seq.), are the primary authorities for EPA's activities in this area.

Other Federal Agencies

Among the other agencies that play a role in food safety, USDA's Agricultural Research Service (ARS) performs food safety research in support of FSIS's inspection program. It has scientists working in animal disease bio-containment laboratories in Plum Island, NY, and Ames, IA. USDA's Animal and Plant Health Inspection Service (APHIS) indirectly protects the nation's food supply through programs to protect plant and animal resources from domestic and foreign pests and diseases, such as brucellosis and bovine spongiform encephalopathy (BSE, or "mad cow" disease). The Department of Homeland Security (DHS) is to coordinate many food security activities, including at U.S. borders.

USDA's Agricultural Marketing Service (AMS) is responsible for establishing quality and marketing grades and standards for many foods (including dairy products, fruits and vegetables, livestock, meat, poultry, seafoods, and shell eggs), and for certifying quality programs and conducting quality grading services. Accordingly, AMS is primarily responsible for ensuring *product quality* and not *food safety*.

Congressional Committees

In the Senate, food safety issues are considered by the Committees on Agriculture, Nutrition, and Forestry; Homeland Security and Governmental Affairs; and Health, Education, Labor, and Pensions. In the House, various food safety activities fall under the jurisdiction of the Committees on Agriculture; Energy and Commerce; Oversight and Government Reform; and Science. Agriculture subcommittees of the House and Senate Appropriations Committees also serve oversight and funding roles in how the major agencies carry out food safety policies.

Table 1. Major Federal Food Safety Agencies

Agency	Major Responsibilities and Activities	Primary Authorities
Department of Health and Human Services		
Food and Drug Administration[a]	Ensuring that all domestic and imported foods, except processed egg products and major types of meat and poultry, are safe, wholesome, and properly labeled, by setting safety and sanitation standards, periodically inspecting manufacturing facilities, reviewing records of and spot-checking imports. Also overseeing the safety of animal drugs and feeds including those used in food-producing animals	As may be amended by the FDA Food Safety Modernization Act (FSMA): Federal Food, Drug, and Cosmetic Act (FFDCA; 21 U.S.C. 301), Public Health Service Act (42 U.S.C. 201), Egg Products Inspection Act (21 U.S.C. 1031), Public Health Security and Bioterrorism Preparedness and Response Act (21 U.S.C. 341), others
Centers for Disease Control and Prevention	Monitoring, identifying, and investigating foodborne diseases; developing and evaluating improved epidemiological and laboratory methods	Public Health Service Act (42 U.S.C. 201)
Department of Agriculture		
Food Safety Inspection Service[a]	Regulating the safety, wholesomeness and proper labeling of most commercial types of both domestic and imported meat and poultry, catfish products, and processed egg products, by approving establishment designs, safety plans; inspecting every animal and carcass in slaughtering plants and daily inspecting all meat and poultry processing plants; determining the equivalency of importing countries' meat and poultry safety systems	Federal Meat Inspection Act (21 U.S.C. 601), Poultry Products Inspection Act (21 U.S.C. 451), Egg Products Inspection Act
Animal and Plant Health Inspection Service	Overseeing animal and plant health, including the prevention of foreign diseases and pests, eradication and containment of such problems domestically (including those that threaten public health)	Animal Health Protection Act (7 U.S.C. 8301), Plant Health Protection Act (7 U.S.C. 7701)
Agricultural Marketing Service	Establishing quality and marketing grades and standards for dairy products, fruits and vegetables, livestock, meat, poultry, seafoods, and shell eggs; certifying quality programs; conducting quality grading services, generally user fee-funded	Agricultural Marketing Act of 1946 (7 U.S.C. 1621), Egg Products Inspection Act (21 U.S.C. 1031), Agricultural Marketing Agreement Act (7 U.S.C. 601)

Table 1. (Continued)

Agency	Major Responsibilities and Activities	Primary Authorities
Food and Nutrition Service	Encouraging and coordinating efforts to ensure the safety of foods in school lunch and other domestic programs	Program subsidies authorized by Richard B. Russell National School Lunch Act (42 U.S.C. 1751),Child Nutrition Act (42 U.S.C. 1771).
Grain Inspection, Packers and Stockyards Administration	Setting quality standards for, and testing, grains and related commodities, primarily for marketing purposes	U.S. Grain Standards Act (7 U.S.C. 71), Agricultural Marketing Act of 1946
Agricultural Research Service	Conducts in-house USDA research on agricultural and food topics, of which food safety is one of many	Numerous laws dating to the Department of Agriculture Organic Act of 1862 (7 U.S.C. 2201 note), up through and including recent omnibus farm laws
Cooperative State Research, Education, and Extension Service	Coordinates and administers federal funding of land grant and other institutions to conduct agricultural and food research, education and extension activities; food safety is one of many subject areas	Numerous laws dating to the Department of Agriculture Organic Act of 1862, up through and including recent omnibus farm laws
Department of Commerce		
National Oceanic and Atmospheric Administration	Offering a variety of voluntary seafood safety and quality inspection services on a fee-for-service basis	Agricultural Marketing Act of 1946, Fish and Wildlife Act of 1956 (16 U.S.C. 742)
Environmental Protection Agency	Regulating pesticide products; setting maximum allowable tolerances for residue levels on food commodities and animal feeds	Federal Insecticide, Fungicide, and Rodenticide Act (7 U.S.C. 136), FFDCA
Federal Trade Commission	Enforcing federal prohibitions against unfair or deceptive acts or practices in trade, including consumer deception regarding foods	Federal Trade Commission Act (15 U.S.C. 41)
Department of the Treasury		
Alcohol and Tobacco Tax and Trade Bureau	Administering and enforcing laws on the production, safety, distribution and use of alcoholic beverages	Federal Alcohol Administration Act (27 U.S.C. 201), Internal Revenue Code (26 U.S.C. Ch. 51)
Department of Homeland Security		
U.S. Customs and Border Protection	Coordinating many food security activities, including at the border; now conducting agricultural border inspection activities formerly done by APHIS	Homeland Security Act (6 U.S.C. 101)

Source: Prepared by CRS based in part on various reports by the Government Accountability Office.

[a] These agencies have the leading food safety regulatory authorities.

Table 2. Selected Comparison of FSIS and FDA Responsibilities

Activity	Food Safety and Inspection Service	Food and Drug Administration (Foods Program only)
Primary Authorizations	Federal Meat Inspection Act (21 U.S.C. 601), Poultry Products Inspection Act (21 U.S.C. 451), Egg Products Inspection Act (21 U.S.C. 1031)	As may be amended by the FDA Food Safety Modernization Act (FSMA): Federal Food, Drug, and Cosmetic Act (FFDCA; 21 U.S.C. 301; Public Health Service Act (42 U.S.C. 201); Egg Products Inspection Act (21 U.S.C. 1031); Public Health Security and Bioterrorism Preparedness and Response Act (21 U.S.C. 341)
Foods Regulated	Major types of domestic and imported meat and poultry and their products; catfish products; processed (dried, frozen, liquid) egg products (20% of at-home U.S. food spending)	All other domestic and imported foods, also animal drugs and feeds including those used in food-producing animals (80% of at-home U.S. food spending)
Funding (enacted FY2012)	Appropriated: $1.004 billion for FY2012. Expected user fees are estimated to include another $150 million.	Appropriated: $866.1 million for FDA's Foods Program, not including funding from expected user fees. Expected user fees are estimated to include another $79 million. Including authorized fees, total available funding is estimated at about $945 million.
Staff (2011)	9,600 FTEs	3,400 FTEs
Domestic facilities	6,300 slaughter and/or processing establishments	68,000 subject to inspection
Inspection Approach	Ante- and post-mortem inspection of every animal, carcass and part; traditionally organoleptic (but see "Food safety plans" below); only USDA-inspected and passed products may enter commerce	Prohibits adulteration or misbranding; relies on facilities that manufacture, process, pack, or hold food for humans or animals to be meet prescribed standards (e.g., regarding additives, contaminants, etc.); all facilities must register, report changes in timely manner.
Required inspection frequency	Slaughter plants: all times of operation; processing plants: at least once daily	FSMA requires increased inspection rates for any registered facility, particularly those identified as "high-risk." Domestic high-risk facilities are to be inspected not less than once in the five-year period after enactment, and not less than once every three years thereafter. Domestic non-high-risk facilities are to be inspected not less than once in the seven-year period after enactment, and not less than once every five years thereafter.

Table 2. (Continued)

Activity	Food Safety and Inspection Service	Food and Drug Administration (Foods Program only)
Food safety plans	Requires all establishment to prepare and have preapproved "HACCP" (hazard analysis and critical control point) plans determining risks, controlling them (with documentation)	Prior to FSMA, facilities followed general regulations on good manufacturing practices (GMPs) to address safe handling and plant sanitation—except a form of HACCP required for seafood, low-acid canned foods, juices. FSMA §103 created new requirements for facilities to evaluate hazards, implement preventive controls, monitor controls, and maintain records. FDA rulemaking is clarifying requirements under new written HACCP-type and/or broader written food safety plans as part of its so-called Hazard Analysis and Risk-Based Preventive Controls.
Imports	Specified products only from countries where FSIS has determined "equivalence" of foreign safety system, with annual verification; imports exempt from prior notice but subject to reinspection at 150 import establishments (est. 10% reinspected)	Prior to FSMA, food safety system equivalence was not determined beforehand; reliance on inspections was at 300 ports (est. 1% of notified entries inspected). FSMA provides for tighter controls and use certification or verification systems for imported foods(to be determined by FDA rulemaking). At least 600 foreign facilities must be inspected the year following enactment, and in each of the subsequent five years the number of foreign facilities inspected is to double.
Third party certification	Private labs accredited for chemical testing of meat and poultry (for imports, see above)	Prior to FSMA, there was no accreditation for food testing labs or use of third parties for import oversight. FSMA §202 requires FDA to establish a program for testing of food by accredited labs and to recognize accreditation bodies to accredit labs. FSMA §303 creates a system of accreditation of third-party auditors and audit agents to certify importing entities. FDA's rulemaking is ongoing.
On-farm oversight	FSIS inspection authority begins at slaughter plant	Prior to FSMA, those engaged solely in harvesting, storing or distributing raw agricultural commodities were generally exempt from registration, GMP regulations, and record-keeping. FSMA §105 created new farm-level requirements, particularly for fresh produce determined to be higher-risk (FDA rulemaking is ongoing). Some small farm businesses are exempt from regulation.

Activity	Food Safety and Inspection Service	Food and Drug Administration (Foods Program only)
Labeling	Review and preapproval required for all labels	All foods must adhere to food labeling requirements such as statement of identity, declaration of net contents, nutrition labeling; labels cannot be false or misleading.
Notification Requirements	P.L. 110-246 §11017 amended meat and poultry laws to require an establishment to notify USDA if it has reason to believe that an adulterated or misbranded product has entered commerce	P.L. 110-85 (amended by FSMA) requires FDA to maintain a reportable food registry for industry to report food safety cases in order to help FDA better track patterns and target inspections. FSMA §204 provided for an enhanced tracing system for foods that FDA determines to pose a higher food safety risk. As part of the ongoing rulemaking process, FDA has launched product tracing pilots.
Recall Authority	No authority to mandate recalls; relies on voluntary efforts	Prior to FSMA, FDA had no authority to mandate recalls (except infant formula). FSMA §206 provides for mandatory recall authority where there is a reasonable probability that a food is adulterated or misbranded, and its use or exposure to it will cause serious adverse health consequences or death. Civil/criminal penalties apply for failure to comply with a recall order.

Source: Prepared by CRS.

End Notes

[1] Roughly two-thirds of the $1 trillion is for domestically produced farm foods; imports and seafood account for the balance. USDA, Economic Research Service (ERS) data, at http://www.ers.usda.gov/Browse/FoodSector/.

[2] U.S. Department of Health and Human Services, Centers for Disease Control and Prevention, "Estimates of Foodborne Illness in the United States," http://www.cdc.gov/foodborneburden/2011-foodborne-estimates.html.

[3] The 20% estimate is based on information reported by the Government Accountability Office (GAO) in "Revamping Oversight of Food Safety," prepared for the 2009 Congressional and Presidential Transition, and appears to represent proportions of total spending for food consumed at home. The 10% estimate is based on data from USDA's Economic Research Service (ERS) on U.S. per capita food consumption at http://www.ers.usda.gov/data/foodconsumption/.

[4] For more information, see CRS Report R40443, *The FDA Food Safety Modernization Act (P.L. 111-353)*. FSMA does not directly address meat and poultry products under USDA's jurisdiction.

[5] CBO, Cost Estimate, "S. 510, Food Safety Modernization Act, as reported by the Senate Committee on Health, Education, Labor, and Pensions on December 18, 2009, incorporating a manager's amendment released on August 12, 2010," August 12, 2010. Reflecting the August 2010 Senate amendment to S. 510.

[6] For more information, see CRS Report R41964, *Agriculture and Related Agencies: FY2012 Appropriations*.

[7] FDA "Operating Plan for FY 2011 and Comparisons to FY 2010," http://www.hhs.gov/asfr/ob/docbudget/ 2011operatingplan_fda.pdf.

[8] HHS OIG, *FDA Inspections of Domestic Food Facilities* (OEI-02-08-00080), Table 1 , April 2010, http://oig.hhs.gov/ oei/reports/oei-02-08-00080.pdf.

[9] HHS, OIG, *Vulnerabilities in FDA's Oversight of State Food Facility Inspections* (OEI-02-09-00430), December 2011, http://oig.hhs.gov/oei/reports/oei-02-09-00430.pdf.

[10] GAO, *Fundamental Changes Needed to Ensure Safe Food* (GAO-02-47T), October 10, 2001, http://www.gao.gov/ new.items/d0247t.pdf.

[11] See footnote 3.

[12] USDA's Agricultural Marketing Service (AMS) also has responsibility for ensuring shell eggs for quality, but not safety, considerations (see Table 1).

[13] USDA, 2012 Explanatory Notes, Food Safety and Inspection Service, February 12, 2011, http://www.obpa.usda.gov/ 21fsis2012notes.pdf.

[14] USDA, "Listing of Participating States," http://www.fsis.usda.gov/regulations_and_policies/ Listing_of_participating_states/index.asp

[15] The 2008 farm bill (P.L. 110-246, § 11017) contained new provisions intended to enable more interstate shipment of state-inspected products.

INDEX

D

E

N

O

P

Q

R

S

T

U

V

W